Modern Infrared Spectroscopy

ANALYTICAL CHEMISTRY BY OPEN LEARNING

ACOL (Analytical Chemistry by Open Learning) is a well established series which comprises 33 open learning books and 9 computer based training packages. This open learning material covers all of the important techniques and fundamental principles of analytical chemistry.

Books

Samples and Standards
Sample Pretreatment
Classical Methods Vols I and II
Measurement, Statistics and Computation
Using Literature
Instrumentation
Chromatographic Separations
Gas Chromatography
High Performance Liquid Chromatography
Electrophoresis
Thin Layer Chromatography
Visible and Ultraviolet Spectroscopy
Fluorescence and Phosphorescence
Atomic Absorption and Emission Spectroscopy
Nuclear Magnetic Resonance Spectroscopy
X-Ray Methods
Mass Spectrometry
Scanning Electron Microscopy and Microanalysis
Principles of Electroanalytical Methods
Potentiometry and Ion Selective Electrodes
Polarography and Voltammetric Methods
Radiochemical Methods
Clinical Specimens
Diagnostic Enzymology
Quantitative Bioassay
Assessment and Control of Biochemical Methods
Thermal Methods
Microprocessor Applications
Chemometrics: Experimental Design
Environmental Analysis
Quality in the Analytical Chemistry Laboratory
Modern Infrared Spectroscopy

Software

Atomic Absorption Spectroscopy
High Performance Liquid Chromatography
Polarography
Radiochemistry
Gas Chromatography
Fluorescence
Quantitative IR and UV
Chromatography
Measure Your Own Quality

Series Editor: David J. Ando

Further information: ACOL Office
The University of Greenwich
Unit 42, Dartford Trade Park
Hawley Road
Dartford
DA1 1PF

Modern Infrared Spectroscopy

Analytical Chemistry by Open Learning

Author:
BARBARA STUART
University of Greenwich

Editor:
DAVID J. ANDO
University of Greenwich

Published on behalf of ACOL (University of Greenwich)
by
JOHN WILEY & SONS
Chichester • New York • Brisbane • Toronto • Singapore

Published by John Wiley & Sons, Ltd,
Baffins Lane, Chichester,
West Sussex PO19 1UD, England

Telephone: National (01243) 779777
International (+44) 1243 779777

Other Wiley Editorial Offices

John Wiley & Sons, Inc., 605 Third Avenue,
New York, NY 10158-0012, USA

Jacaranda Wiley Ltd, 33 Park Road, Milton,
Queensland 4064, Australia

John Wiley & Sons (Canada) Ltd, 22 Worcester Road,
Rexdale, Ontario M9W 1L1, Canada

John Wiley & Sons (SEA) Pte Ltd, 37 Jalan Pemimpin #05-04,
Block B, Union Industrial Building, Singapore 2057

British Library Cataloguing in Publication Data

A catalogue record for this book is available from the British Library

ISBN 0 471 95916 2 (cloth)
ISBN 0 471 95917 0 (paper)

Typeset in 11/13pt Times by Mackreth Media Services, Hemel Hempstead, Herts
Printed and bound in Great Britain by Biddles Ltd, Guildford, Surrey
This book is printed on acid-free paper responsibly manufactured from sustainable forestation, for which at least two trees are planted for each one used for paper production.

THE UNIVERSITY OF GREENWICH ACOL PROJECT

This series of easy-to-read books has been written by some of the foremost lecturers in Analytical Chemistry in the United Kingdom. These books are designed for training, continuing education and updating of all technical staff concerned with Analytical Chemistry.

These books are for those interested in Analytical Chemistry and instrumental techniques who wish to study in a more flexible way than traditional institute attendance, or to augment such attendance.

ACOL also supply a range of training packages which contain computer software together with the relevant ACOL book(s). The software teaches competence in the laboratory by providing experience of decision making in the laboratory, often based on the simulation of instrumental output, while the books cover the requisite underpinning knowledge.

The Royal Society of Chemistry uses ACOL material to run regular series of courses based on distance learning and regular workshops.

Further information on all ACOL materials and courses may be obtained from:

ACOL Office, University of Greenwich, Unit 42, Dartford Trade Park, Hawley Road, Dartford, DA1 1PF. Tel: 0181-331-9600, Fax: 0181-331-9672.

How to Use an Open Learning Book

Open Learning books are designed as a convenient and flexible way of studying for people who, for a variety of reasons, cannot use conventional education courses. You will learn from this book the principles of one subject in Analytical Chemistry, but only by putting this knowledge into practice, under professional supervision, will you gain a full understanding of the analytical techniques described.

To achieve the full benefit from an open learning text you need carefully to plan your place and time of study.

- Find the most suitable place to study where you can work without disturbance.

- If you have a tutor supervising your study discuss with this person the date by which you should have completed this text.

- Some people study perfectly well in irregular bursts; however, most students find that setting aside a certain number of hours each day is the most satisfactory method. It is for you to decide which pattern of study suits you best.

- If you decide to study for several hours at once, take short breaks of five or ten minutes every half hour or so. You will find that this method maintains a higher overall level of concentration.

Before you begin a detailed reading of this book, familiarise yourself with the general layout of the material. Have a look at the course contents list at the front of the book and flip through the pages to get a general impression of the way the subject is dealt with. You will find that there is space on the pages to make comments alongside the text as you study — your own notes for highlighting points that you feel are particularly important. Indicate in the margin the points you would like to discuss further with a tutor or fellow student. When you come to revise, these personal study notes will be very useful.

Π When you find a paragraph in the text marked with a symbol such as is shown here, this is where you get involved. At this point you are directed to do things: draw graphs, answer questions, perform calculations, etc. Do make an attempt at these activities. If necessary, cover the succeeding response with a piece of paper until you are ready to read on. This is an opportunity for you to learn by participating in the subject, and although the text continues by discussing your response, there is no better way to learn than by working things out for yourself.

We have introduced self-assessment questions (SAQs) at appropriate places in the text. These SAQs provide for you a way of finding out if you understand what you have just been studying. There is space on the page for your answer and for any comments you want to add after reading the author's response. You will find the author's response to each SAQ at the end of the book. Compare what you have written with the response provided and read the discussion and advice.

At intervals in the text you will find a Summary and a list of Objectives. The Summary will emphasise the important points covered by the material you have just read, while the Objectives will give you a checklist of tasks you should then be able to achieve.

You can revise the book, perhaps for a formal examination, by re-reading the Summary and the Objectives, and by working through some of the SAQs. This should quickly alert you to areas of the text that need further study.

At the end of this book you will find, for reference, lists of commonly used scientific symbols and values, units of measurement, and also a periodic table.

Contents

Study Guide

The first Infrared Spectroscopy Unit in the ACOL series was written by Bill George and Peter McIntyre. This successor to that Unit has been written by Barbara Stuart, who is currently lecturing at the Univesity of Technology, Sydney, Australia.

This present text has been designed to provide you with a working knowledge of modern infrared spectroscopy and how to apply the technique to a range of analytical and structural problems.

It will be assumed that you have an understanding of chemistry equivalent to that of a student who has passed HNC or HTC in chemistry (BTEC), and a knowledge of physics up to at least GCSE level. However, there are many spectroscopists who have no formal chemical education.

Because infrared spectroscopy can be applied to a wide range of disciplines, you are bound to find that there are some topics that you would like to study in more detail than is given in this text. Suitable books that you could use as a starting point are listed in the Bibliography. Likewise, you may find that some of the specialised topics covered are not particularly relevant to your work — you may like to treat those topics as optional. This may particularly be the case when you reach Chapter 6.

You will find that the best way to learn is through hands-on experience. Like most analytical techniques, knowledge comes through many hours of standing in front of an instrument and making many mistakes. This can be frustrating, but it is probably the best way to learn to recognise the problems which can arise in infrared spectroscopy. Hopefully, this text will help to minimise the amount of mistakes made and the time taken to complete your experiments!

Supporting Practical Work

1. GENERAL CONSIDERATIONS

Most analytical laboratories are equipped with an infrared spectrometer, be it an older style dispersive machine or a more modern Fourier-transform instrument. The technique is typically used in conjunction with a variety of other analytical methods, such as nuclear magnetic resonance spectroscopy, mass spectrometry, ultraviolet–visible spectroscopy, or chromatography, in order to obtain information about a wide range of samples.

The operation of dispersive instruments is generally straightforward, with most of the skill involved being associated with the sample preparation. If you have access to a Fourier-transform infrared spectrometer, you will probably need to spend some time becoming familiar with the computer software which drives the instrument. Some suggestions for experiments to try out for yourself are listed here. It is not necessary to attempt all of those mentioned. You may decide to concentrate on samples relevant to your field of interest.

2. AIMS

(a) To provide practical experience in the use of an infrared spectrometer.

(b) To provide experience in various methods of infrared sample preparation.

(c) To demonstrate the feasibility of qualitative and quantitative analysis.

(d) To illustrate some of the important theoretical principles of infrared spectroscopy.

3. SUGGESTED EXPERIMENTS

(a) The examination of the spectra of benzamide powder, both as a potassium bromide disc and as a Nujol mull.

(b) The examination of the spectrum of a film of polystyrene cast from chloroform.

(c) The quantitative analysis of a multicomponent mixture of xylene liquids.

(d) The quantitative analysis of glucose in aqueous solution using attenuated total reflectance spectroscopy.

Bibliography

There are numerous books available which detail the theory and the many applications of infrared spectroscopy. Some of these are listed here.

GENERAL INTRODUCTION TO INFRARED SPECTROSCOPY

P. W. Griffiths, *Chemical Infrared Fourier Transform Spectroscopy*, Wiley, 1975.

ORGANIC COMPOUNDS

W. Kemp, *Organic Spectroscopy*, 3rd Edn, Macmillan, 1991.
N. B. Coltrup, L. H. Daly and S. E. Wiberley, *Introduction to Infrared and Raman Spectroscopy*, 3rd Edn, Academic Press, 1990.

D. Lin-Vien, N. B. Coltrup, W. G. Fateley and J. G. Grasselli, *The Handbook of Infrared Frequencies of Organic Molecules*, Academic Press, 1991.

INORGANIC COMPOUNDS

K. Nakamoto, *Infrared and Raman Spectra of Inorganic and Coordination Compounds*, 4th Edn, Wiley, 1986.

N. B. Coltrup, L. H. Daly and S. E. Wiberley, *Introduction to Infrared and Raman Spectroscopy*, 3rd Edn, Academic Press, 1990.

L. C. Thomas, *Interpretation of the Infrared Spectra of Organophosphorus Compounds*, Heyden and Son, 1974.

POLYMERS

D. I. Bower and W. F. Maddams, *The Vibrational Spectroscopy of Polymers*, Cambridge University Press, 1989.

H. W. Siesler and K. Holland-Moritz, *Infrared and Raman Spectroscopy of Polymers*, Marcel Dekker, 1980.

BIOLOGICAL SYSTEMS

R. E. Hester and R. B. Girling (Eds), *Spectroscopy of Biological Molecules*, Royal Society of Chemistry, 1991.

SURFACE CHEMISTRY

R. J. H. Clark and R. E. Hester (Eds), *Spectroscopy of Surfaces*, Wiley, 1988.

J. W. Niemantsverdriet, *Spectroscopy in Catalysis: An Introduction*, VCH, 1993.

VARIOUS APPLICATIONS

J. R. Durig (Ed.), *Chemical, Biological and Industrial Applications of Infrared Spectroscopy*, Wiley, 1985.

Acknowledgements

Figures 1.2a, 1.2b, 1.2c and 2.3a are redrawn from E. F. H. Brittain, W. O. George and C. H. J. Wells, *Introduction to Molecular Spectroscopy*, Academic Press, 1975. Permission has been requested.

Figure 1.3a is redrawn from G. Herzberg, *Molecular Spectra of Polyatomic Molecules*, Van Nostrand, 1954. Permission has been requested.

Figures 2.4c and 2.4d are redrawn from A. J. Barnes and W. J. Orville-Thomas (Eds), *Vibrational Spectroscopy — Modern Trends*, Elsevier, 1977. Permission has been requested.

Figures 3.2c, 3.3f and 3.4b are redrawn from a Perkin-Elmer sales leaflet. Permission has been requested.

Figure 3.3c is redrawn from R. R. Hill and D. A. E. Rendell, *The Interpretation of Infrared Spectra: A Programmed Introduction*, Heyden and Son Ltd, 1975. Permission has been requested.

Figures 5.2c and 5.2d are redrawn from the Perkin-Elmer *Application Note on Derivative Spectroscopy*, Perkin Elmer, 1984. Permission has been requested.

Figures 6.3 and 7.3d are redrawn from J. Emsley and D. Hall, *The Chemistry of Phosphorus*, Harper and Row, 1976. Permission has been requested.

Figure 6.4b is redrawn from W. Klopffer, *Introduction to Polymer Spectroscopy*, Springer-Verlag, 1984. Permission has been requested.

Figure 6.6 is redrawn from K. S. Kalasinsky, B. Levine, M. L. Smith, J. Magluilo and T. Schaefer, *J. Anal. Toxicol.* **17**, 359 (1993). Permission has been requested.

Figure 6.9 is redrawn from J. R. Durig (Ed.), *Chemical, Biological and Industrial Applications of Infrared Spectroscopy*, Wiley, 1985. Permission has been requested.

Figure 7.4e is redrawn from S. Pawlenko, *Organosilicon Chemistry*, Walter de Gruyer, 1986. Permission has been requested.

1. Introduction

1.1. GENERAL INTRODUCTION

Infrared spectroscopy is certainly one of the most important analytical techniques available to today's chemists. Infrared spectrometers have been commercially available since the 1940s. At that time the instruments relied on prisms to act as dispersive elements, but by the mid 1950s, diffraction gratings had been introduced into dispersive machines. The most significant advances in infrared spectroscopy, however, have come about with the introduction of Fourier-transform spectrometers. This type of instrument employs an interferometer and exploits the well established mathematical process of Fourier transformation. Fourier-transform infrared (FT-IR) spectroscopy has dramatically improved the quality of infrared spectra and minimised the time required to obtain data. Also, with improvements to computers in recent years, infrared spectroscopy has made great strides.

One of the great advantages of infrared spectroscopy is that virtually any sample in virtually any state can be studied. Liquids, solutions, pastes, powders, films, fibres, gases and surfaces can all be examined by a judicious choice of sampling technique. As a consequence of the improved instrumentation, a variety of new sensitive techniques have been developed in order to examine formerly intractable samples. Polymers, drugs, organic compounds, inorganic samples, biological samples, paints, oils, lubricants, fibres, catalysts, minerals, organometallics, coal, forensic samples, clays, atmospheric samples and food additives have all been studied by using infrared spectroscopy.

Infrared spectroscopy is a technique based on the vibrations of the atoms of a molecule. An infrared spectrum is obtained by passing radiation through a sample and determining what fraction of the incident radiation is absorbed at a particular energy. The energy at which any peak in an absorption spectrum appears corresponds to the frequency of a vibration of a part of a sample molecule. In this introductory chapter we will discuss the basic ideas and definitions associated with infrared spectroscopy. We will look in some detail at the vibrations of molecules as these are crucial to the interpretation of infrared spectra.

Once you have worked through this chapter you will have some idea about the information to be gained from infrared spectroscopy. The following chapter will help you understand how an infrared spectrometer produces a spectrum. After working through that chapter you will probably be ready to record a spectrum of your own and to do that you need to decide on an appropriate sampling technique. The sampling procedure depends very much on the type of sample you want to examine, for example, whether it is a solid, liquid or gas. Chapter 3 outlines the various sampling techniques that are commonly available. Once you have recorded your spectrum, you obviously need to be able to extract the information it can provide. Chapter 4, on spectrum interpretation, will help you interpret the information to be gained from an infrared spectrum. As infrared spectroscopy is now used in such a wide variety of scientific fields, some of the many applications of the technique are examined in Chapter 6. This chapter should provide you with some guidance as to how to approach a particular analytical problem in your field. Finally, to give you some feel for the interpretation of spectra, Chapter 7 provides some practice examples.

1.2. ELECTROMAGNETIC RADIATION

You will be familiar with the visible part of the electromagnetic spectrum. This radiation is, by definition, visible to the human eye and is also continuous and polychromatic. Other detection systems can reveal radiation which is beyond the visible regions of the spectrum, and these are classified as γ-rays, X-rays, ultraviolet, infrared, microwaves and radiowaves. These regions are shown in Figure 1.2a, together with the processes involved in the interaction of the radiation of these regions with matter. The electromagnetic spectrum and the varied interactions between these radiations and many forms of

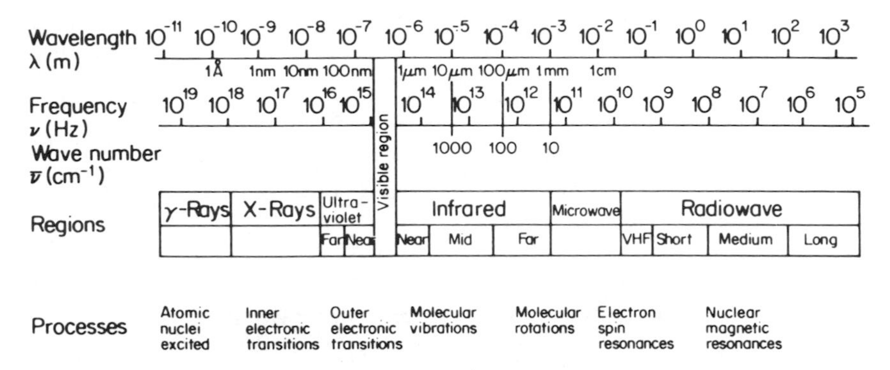

Fig. 1.2a. *The electromagnetic spectrum*

matter can either be considered in terms of classical or quantum theories.

The nature of the various radiations shown in Figure 1.2a have been interpreted by Maxwell's classical theory of electro- and magnetodynamics, hence the term electromagnetic radiation. According to this theory radiation is considered as two mutually perpendicular electric and magnetic fields, oscillating in single planes at right angles to each other. These fields are in phase and are being propagated as a sine wave, as shown in Figure 1.2b. The electric and magnetic vectors are represented by E and B, respectively.

A significant discovery made about electromagnetic radiation was that the velocity of propagation in a vacuum was constant for all regions of the spectrum. This is known as the velocity of light (c) and has the following value:

$$c = 2.997925 \times 10^8 \text{ m s}^{-1}$$

If you can visualise one complete wave travelling a fixed distance each cycle you should be able to see that the velocity of this wave is the product of the *wavelength* λ (the distance between adjacent peaks) and the *frequency* ν (the number of cycles per second). It follows that

$$c = \lambda\nu \qquad (1.1)$$

The presentation of spectral regions may be in terms of wavelength, either as metres or sub-multiples of a metre. The following units are commonly used in spectroscopy:

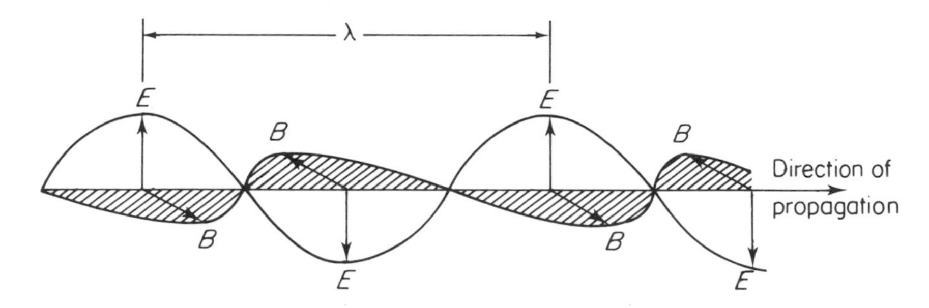

Fig. 1.2b. *An electromagnetic wave*

$$1\,\text{Å} = 10^{-10}\,\text{m};\; 1\,\text{nm} = 10^{-9}\,\text{m};\; 1\,\mu\text{m} = 10^{-6}\,\text{m}$$

Another unit which is commonly used in infrared spectroscopy is the *wavenumber*, which is expressed in cm^{-1}. This is the number of waves in a length of one centimetre and is given by the following relationship:

$$\bar{\nu} = \frac{1}{\lambda} = \frac{\nu}{c} \qquad (1.2)$$

This unit has the advantage of being linear with energy.

During the 19th century a number of experimental observations were made which were not consistent with the classical view that matter could interact with energy in a continuous form. Work by Einstein, Planck and Bohr indicated that in many ways electromagnetic radiation could be regarded as a stream of particles or quanta for which the energy, E, is given by the Bohr equation, as follows:

$$E = h\nu \qquad (1.3)$$

where h is the Planck constant ($h = 6.626 \times 10^{-34}$ J s) and ν is equivalent to the classical frequency.

Processes of electronic change, including those of vibration and rotation associated with infrared spectroscopy, can be represented in terms of quantised discrete energy levels E_0, E_1, E_2, etc., as shown in Figure 1.2c.

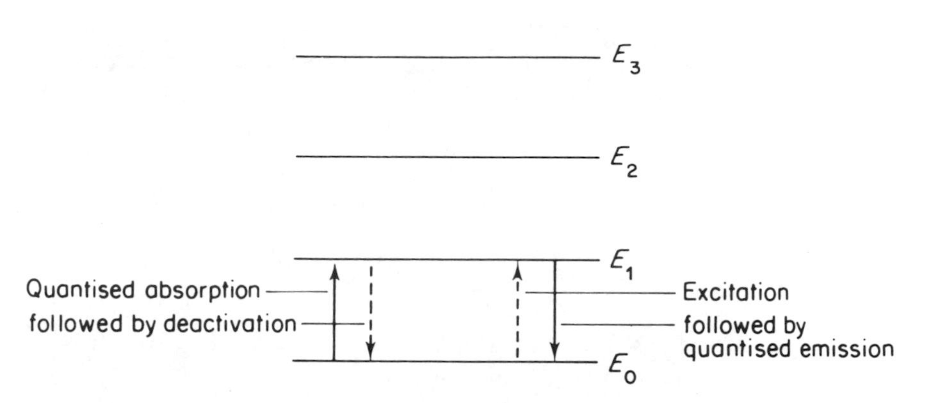

Fig. 1.2c. *Energy levels*

Each atom or molecule in a system must exist in one or other of these levels. In a large assembly of molecules there will be a distribution of all atoms or molecules among these various energy levels.

These energy levels are a function of an integer (the *quantum number*) and a parameter associated with the particular atomic or molecular process associated with that state. Whenever a molecule interacts with radiation, a quantum of energy (or photon) is either emitted or absorbed. In each case the energy of the quantum of radiation must exactly fit the energy gap $E_1 - E_0$, $E_2 - E_1$, etc. The energy of the quantum is related to the frequency by the following:

$$\Delta E = h\nu \tag{1.4}$$

Hence the frequency of emission or absorption of radiation for a transition between the energy states E_0 and E_1 is given by the following relationship:

$$\nu = \frac{E_1 - E_0}{h} \tag{1.5}$$

Associated with the uptake of energy of quantised absorption is some deactivation mechanism whereby the atom or molecule returns to its original state. Associated with the loss of energy by emission of a quantum of energy or photon is some prior excitation mechanism. Both these associated mechanisms are represented by the dotted lines shown in Figure 1.2c.

SAQ 1.2

The molecule hydrogen chloride (HCl) absorbs infrared radiation at 2881 cm^{-1}. Calculate the following:

(i) the wavelength of this radiation;

(ii) the frequency of this radiation;

(iii) the energy change associated with this absorption.

1.3. VIBRATIONS OF MOLECULES

We can understand the interactions of infrared radiation with matter in terms of changes in the molecular dipoles associated with vibrations and rotations. We will not go into great detail about the classical and quantum theories of infrared spectroscopy — such detail is really beyond the scope of this book. Those interested in gaining a more in-depth knowledge of the background theory will find that most standard Physical Chemistry texts provide a detailed coverage of this topic.

1.3.1. Normal Modes of Vibration

A molecule can be looked upon as a system of masses joined by bonds with spring-like properties. First, we will look at the simple case of a diatomic molecule. Diatomic molecules have three degrees of translational freedom and two degrees of rotational freedom. The atoms in molecules can also move relative to one other, i.e. bond lengths can vary or one atom can move out of its present plane. This is a description of the stretching and bending movements which are collectively referred to as *vibrations*. For a diatomic molecule only one vibration is possible, which corresponds to the stretching and compression of the bond. This accounts for one degree of vibrational freedom.

Polyatomic molecules containing many (N) atoms will have $3N$ degrees of freedom. We can distinguish two groups of triatomic molecules, linear and non-linear, with two simple examples being CO_2 and H_2O.

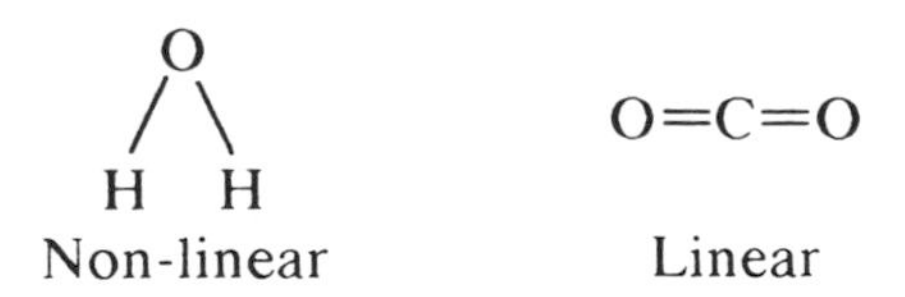

Both CO_2 and H_2O have three degrees of translational freedom. Water has three degrees of rotational freedom, but the linear molecule CO_2 has only two, since no detectable energy is involved in rotation around the O—C—O axis. Subtracting these from $3N$ we have $3N - 5$ for CO_2 (or any linear molecule) and $3N - 6$ for water (or any non-linear molecule). In both examples, N is three, so CO_2 has four vibrational modes and water has three. The degrees of freedom for polyatomic molecules are summarised in Table 1.3.

Table 1.3. *Degrees of freedom for polyatomic molecules*

Degrees of freedom	Linear	Non-linear
Translational	3	3
Rotational	2	3
Vibrational	$3N-5$	$3N-6$
Total	$3N$	$3N$

SAQ 1.3a

How many vibrational degrees of freedom do the following molecules possess?

(i) Methane (CH_4)

(ii) Ethyne ($HC{\equiv}CH$)

Whereas a diatomic molecule has only one mode of vibration which corresponds to a stretching motion, a non-linear B—A—B type triatomic molecule has three modes, two of which correspond to stretching motions and the remainder corresponding to a bending motion. A linear-type triatomic molecule has four modes, two of which have the same frequency and are said to be *degenerate*.

Two other concepts are also used to explain the frequency of vibrational modes. These are the stiffness of the bond and the masses of the atoms at each end of the bond. The stiffness of the bond can be characterised by a proportionality constant termed the *force constant* (k) (derived from Hooke's Law). The *reduced mass* (μ) provides a useful way of simplifying our calculations by combining the individual atomic masses, and may be expressed as follows:

$$\frac{1}{\mu} = \frac{1}{m_1} + \frac{1}{m_2} \tag{1.6}$$

where m_1 and m_2 are the masses of the atoms at the ends of the bond. A practical alternative way of expressing the reduced mass is by the following relationship:

$$\mu = \frac{m_1 m_2}{m_1 + m_2} \tag{1.7}$$

The equation relating force constant, reduced mass and the frequency of absorption is as follows:

$$\nu = \frac{1}{2\pi}\sqrt{\frac{k}{\mu}} \tag{1.8}$$

This equation can be modified so that we can make direct use of the wavenumber values for bond vibrational frequencies:

$$\bar{\nu} = \frac{1}{2\pi c}\sqrt{\frac{k}{\mu}} \tag{1.9}$$

where c is the speed of light.

A molecule can only absorb radiation when the incoming infrared radiation is of the same frequency as one of the fundamental modes of vibration of the molecule. This means that the vibrational motion of a small part of the molecule is increased while the rest of the molecule is left unaffected.

SAQ 1.3b

> Given that the C—H stretching vibration for chloroform occurs at 3000 cm^{-1}, calculate the C—D stretching frequency for deuterochloroform.

Vibrations can involve either a change in bond length (stretching) or bond angle (bending):

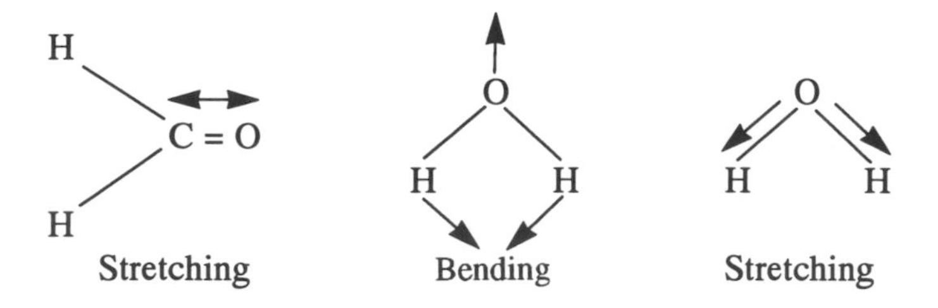

Some bonds can stretch in-phase (*symmetrical* stretching) or out-of-phase (*asymmetric* stretching):

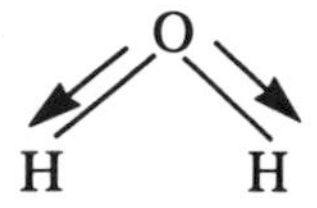

Symmetric stretching

This is *symmetrical* stretching, while that for an alcohol (R—O—H) is *asymmetric* stretching.

Asymmetric stretching

If a molecule has different terminal atoms, such as HCN, ClCN or ONCl, then the two stretching modes are no longer symmetric and asymmetric vibrations of similar bonds, but will have varying proportions of the stretching motions of each group. In other words, the amount of *coupling* will vary. This effect is shown in Figure 1.3a for HCN and ClCN.

We will also look at bending vibrations by using the molecule bromochloromethane (H_2CBrCl) as an example. It is best to consider the molecule being cut by a plane through the hydrogen atoms and the carbon atom. The hydrogens can move in the same direction or in opposite directions in this plane (here the plane of the page).

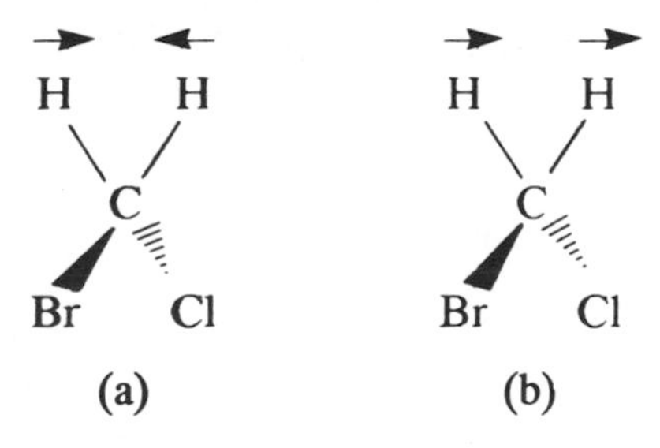

Case (a) is usually referred to as a deformation and case (b) as a rock. There are two more bending vibrations out of this plane. They are referred to as (c) wags and (d) twists:

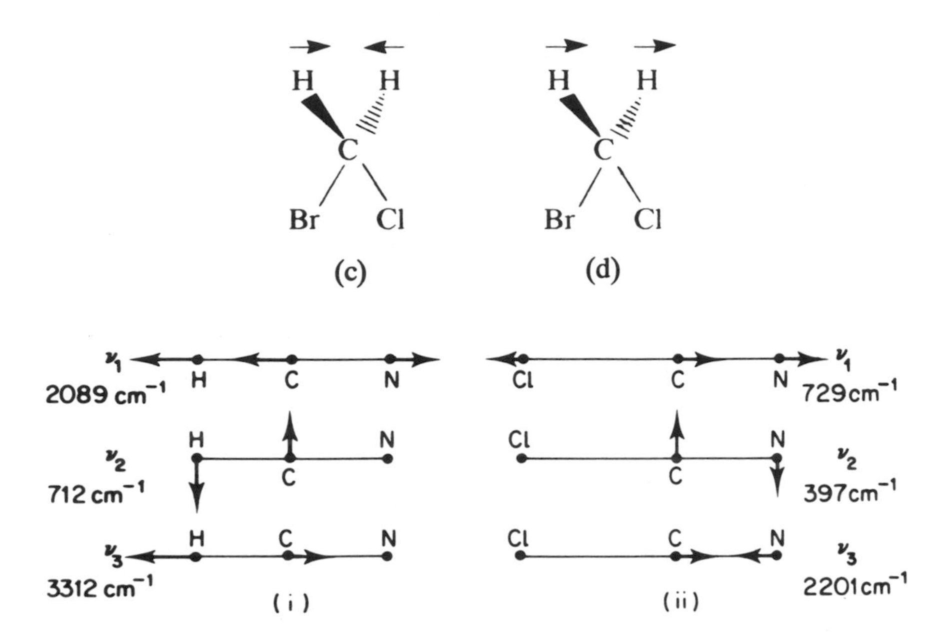

Fig. 1.3a. *Normal vibrations of (i) HCN and (ii) ClCN*

These four vibrations can be simplified to the following:

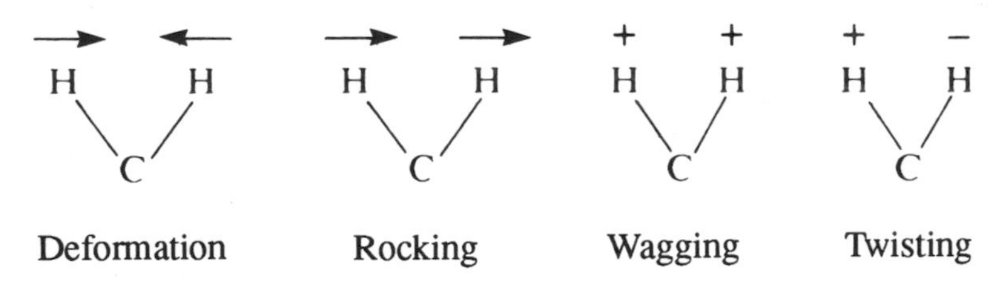

For more complex molecules the analysis becomes simpler, since we can often consider hydrogen atoms in isolation because they are usually attached to more massive, and therefore, more rigid parts of the molecule:

$(CH_3)_2C{=}C(CH_3)H$ (H marked +)

Out-of-plane bending

$(CH_3)_2C{=}C(CH_3)H$ (H with arrow)

In-plane bending

However, there will be many different vibrations for even fairly simple organic molecules. Let us look at the allyl bromide molecule, $CH_2{=}CHCH_2Br$. Applying the formula for this non-linear molecule there are $3N - 6 = 21$ different vibrations. These are C=C stretching, C—H stretching, C—C stretching, C—Br stretching, in-plane and out-of-plane bending of the terminal CH_2 group, in-plane and out-of-plane bending of the =C—H group, deformation vibrations of the CH_2Br group, rocks and wags of the same group, twisting vibrations about the C=C and C—C bonds and bends of the C—C—C skeleton. So you are now probably beginning to appreciate that, even for relatively simple molecules, there are lots of absorption peaks in the infrared spectrum.

The complexity of an infrared spectrum arises from the coupling of vibrations over a large part of, or over, the complete molecule. Such vibrations are called *skeletal* vibrations. Bands associated with skeletal vibrations are likely to conform to a pattern or *fingerprint* of the molecule as a whole, rather than a specific group within the molecule.

1.3.2. Intensity of Infrared Bands

For a vibration to give rise to absorption of infrared radiation, it must cause a change in the dipole moment of the molecule. The larger this change, then the more intense the absorption band will be.

Because of the difference in electronegativity between carbon and oxygen, the carbonyl group is permanently polarised:

$$\gt C^{\delta+}{=}O^{\delta-}$$

Stretching this bond will increase the dipole moment and hence C=O stretching is an intense absorption in acids, ketones, aldehydes, acid chlorides, etc.

In CO_2, two different stretching vibrations are possible, i.e. symmetric and asymmetric.

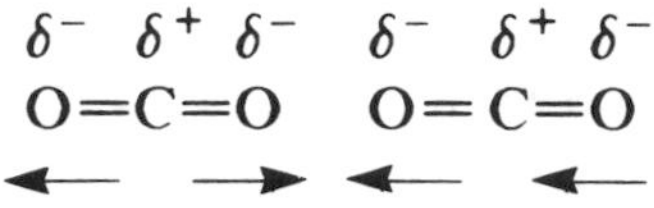

$$\overset{\delta^-}{\mathrm{O}}=\overset{\delta^+}{\mathrm{C}}=\overset{\delta^-}{\mathrm{O}} \qquad \overset{\delta^-}{\mathrm{O}}=\overset{\delta^+}{\mathrm{C}}=\overset{\delta^-}{\mathrm{O}}$$

$$\leftarrow \quad \rightarrow \qquad \leftarrow \quad \leftarrow$$

(a) (b)

Π Which one of the above vibrations is infrared inactive?

A dipole moment is a vector sum, so therefore CO_2 in the ground state has no dipole moment. If the two C=O bonds are stretched symmetrically there is still no net dipole and so there is no infrared activity. However, in the asymmetric stretch the two C=O bonds are now of different length and hence the molecule has a dipole. Therefore, the vibration in (b) above is infrared active.

In practice this black and white situation does not prevail. The change in dipole may be very small and hence lead to a very weak absorption. For example, the double bond in propene (CH_3—CH=CH_2) is only weakly polarised as a result of the inductive effect of the methyl group, with the result that the C=C stretching vibration is almost non-observable.

Bending vibrations provide further difficulties as you now need to think in three dimensions.

Π Consider the symmetrical bending vibration of CO_2:

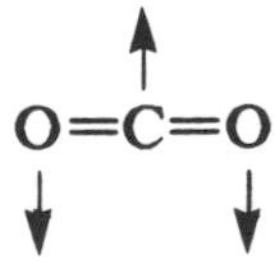

$$\begin{array}{ccc} & \uparrow & \\ \mathrm{O} & = \mathrm{C} = & \mathrm{O} \\ \downarrow & & \downarrow \end{array}$$

Will this vibration be active in the infrared?

The above CO_2 molecule has no dipole moment, but in the bent molecule, i.e.

there is a net dipole in the direction shown. Hence the vibration is infrared active.

SAQ 1.3c

Are the following bending vibrations active or inactive?

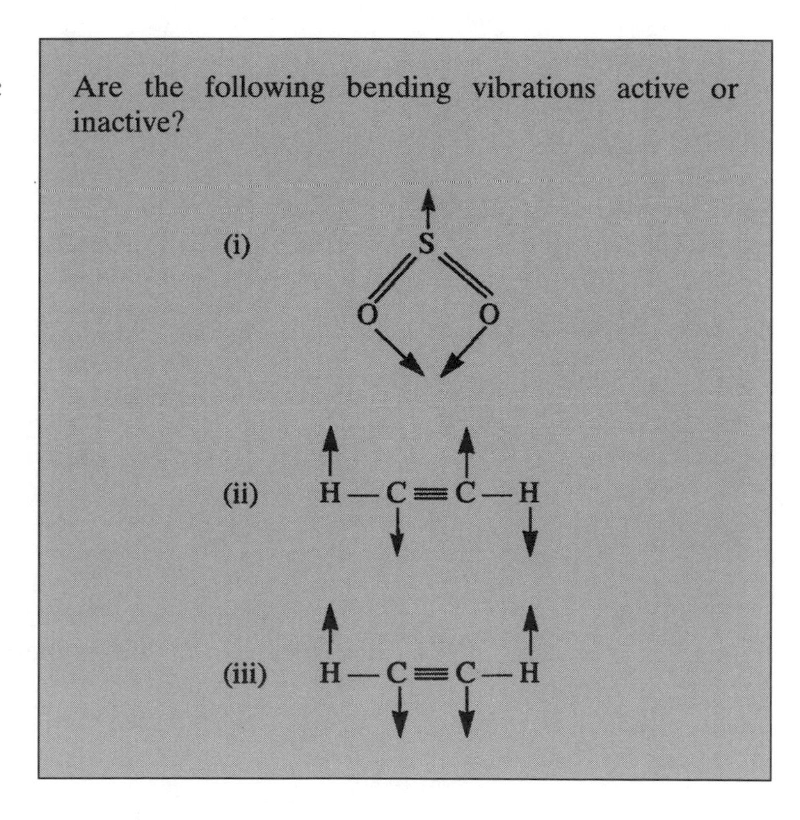

SAQ 1.3c

Consider the C—H out-of-plane bending vibrations below:

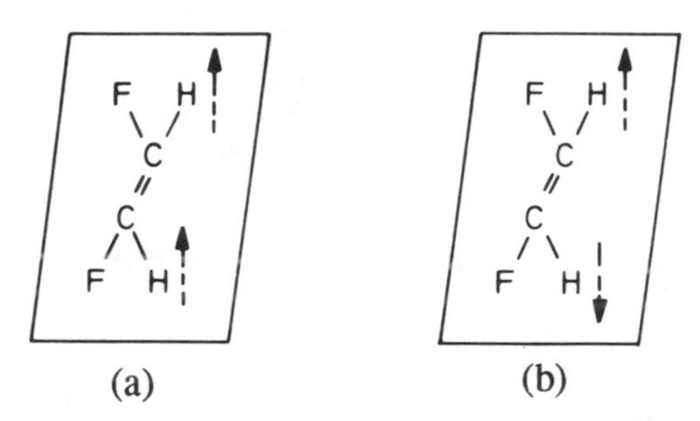

(a) (b)

In (a) the change in dipole moment is in the same direction for both bonds and is additive. In (b) the change in dipole moment is in the opposite direction and is cancelled out. It follows that (a) is infrared active and (b) is infrared inactive. The infrared activity of such vibrations can also be determined by symmetry arguments.

Similar conclusions can be reached for benzene derivatives. Consider the three compounds below:

There are obviously bending and stretching vibrations of the C—H bonds in these molecules which are infrared active. They would be complex to analyse, but the important point is that they will be different for the three different substitution patterns and for many other patterns of more or less highly substituted derivatives.

Symmetrical molecules will have fewer infrared active vibrations than unsymmetrical molecules. This leads to the conclusion that symmetric vibrations will generally be weaker than asymmetric vibrations, since the former will not lead to a change in dipole moment. It follows that the bending or stretching of bonds involving atoms in widely separated groups of the periodic table will lead to intense bands. Vibrations of bonds such as C—C or N═N will give weak bands. This again is because of the small change in dipole moment associated with their vibrations.

Summary

This chapter introduced the ideas which are fundamental to the understanding of infrared spectroscopy.

Initially, the electromagnetic spectrum was considered in terms of various atomic and molecular processes and classical and quantum ideas were introduced.

We then turned to the vibrations of molecules and how they produce an infrared spectrum. The various factors which are responsible for the position and intensity of infrared modes were discussed.

Objectives

On completion of this chapter you should be able to:

- understand the origin of electromagnetic radiation;
- determine the frequency, wavelength, wavenumber and energy change associated with an infrared transition;
- predict the number of fundamental modes of vibration of a molecule;

- appreciate the different possible modes of vibration, i.e. symmetric and asymmetric stretching, bending, degeneracy, coupling and skeletal;

- understand the influence of force constants and reduced masses on the frequency of band vibrations;

- appreciate the factors governing the intensity of bands in an infrared spectrum.

2. Instrumentation

2.1. INTRODUCTION

In this chapter we will look at the instrumentation required to obtain an infrared spectrum. First, we will examine some of the basic features of an infrared spectrum. Traditionally, dispersive instruments, available since the 1940s, have been used. We will look at this type of instrumentation, as dispersive machines are still used in laboratories today. In recent decades, however, a very different method of obtaining an infrared spectrum has increasingly found favour. Fourier-transform infrared (FT-IR) spectrometers are now widely used and have improved the acquisition of infrared spectra dramatically. We will also examine how this type of instrument produces spectra.

2.2. REPRESENTATION OF SPECTRA

Early infrared instruments recorded percentage transmittance over a linear wavelength range. It is now unusual to use wavelength for routine samples and inverse wavelength units are used. This is the wavenumber scale and the units used are cm^{-1}. The output from the instrument is referred to as a *spectrum*. Most commercial instruments present the spectrum with wavenumber decreasing from left to right.

The infrared spectrum can be divided into three regions: the *far infrared* (400–0 cm^{-1}), the *mid infrared* (4000–400 cm^{-1}) and the *near infrared* (14285–4000 cm^{-1}). Most infrared applications employ the mid-infrared region, but the near- and far-infrared regions can also provide information about certain materials (for example, lattice vibrations). The majority of instruments are set up to scan only the mid-infrared range and this is the region that we will concentrate on in later chapters.

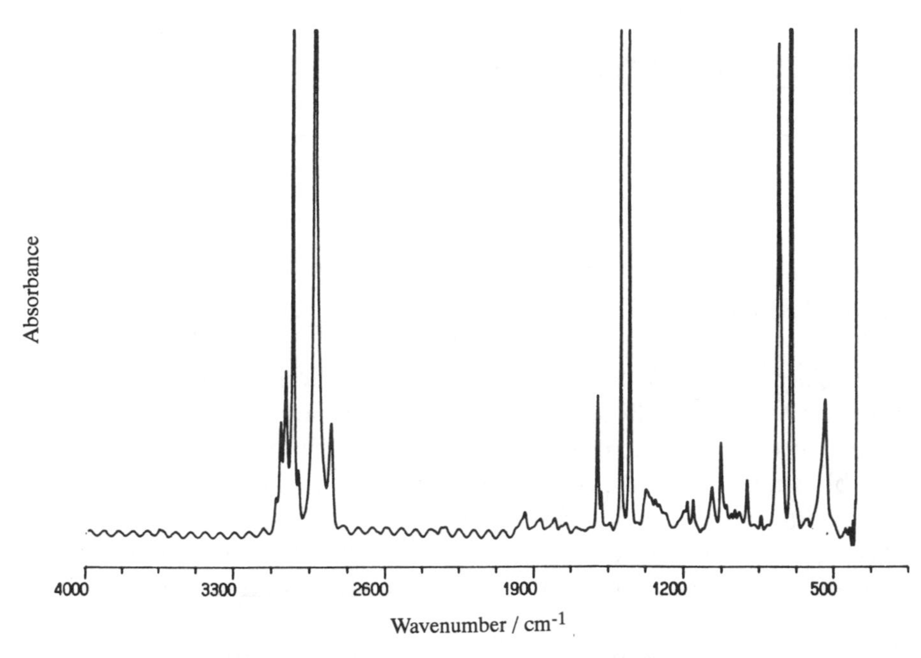

Fig. 2.2a. *An absorbance spectrum of polystyrene*

Generally, there are few infrared bands in the region 4000–1800 cm^{-1}, with many bands between 1800 and 400 cm^{-1}. Often the scale is changed so that the region between 4000 and 1800 cm^{-1} is contracted and the region between 1800 and 400 cm^{-1} is expanded to emphasise the features of interest. The ordinate scale may be presented in % transmittance with 100% at the top of the chart. It is commonplace to have the choice of absorbance or transmittance as a measure of band intensity. We will look at the relationship between these two phenomena in more detail in Chapter 5. Figures 2.2a and 2.2b illustrate the difference in appearance between absorbance and transmittance spectra. It almost comes down to personal preference which of the two modes to use, but the transmittance is traditionally used for spectral interpretation, while absorbance is used for quantitative work.

2.3. DISPERSIVE SPECTROMETERS

The first dispersive infrared instruments employed prisms made of materials such as sodium chloride (NaCl). The popularity of prism

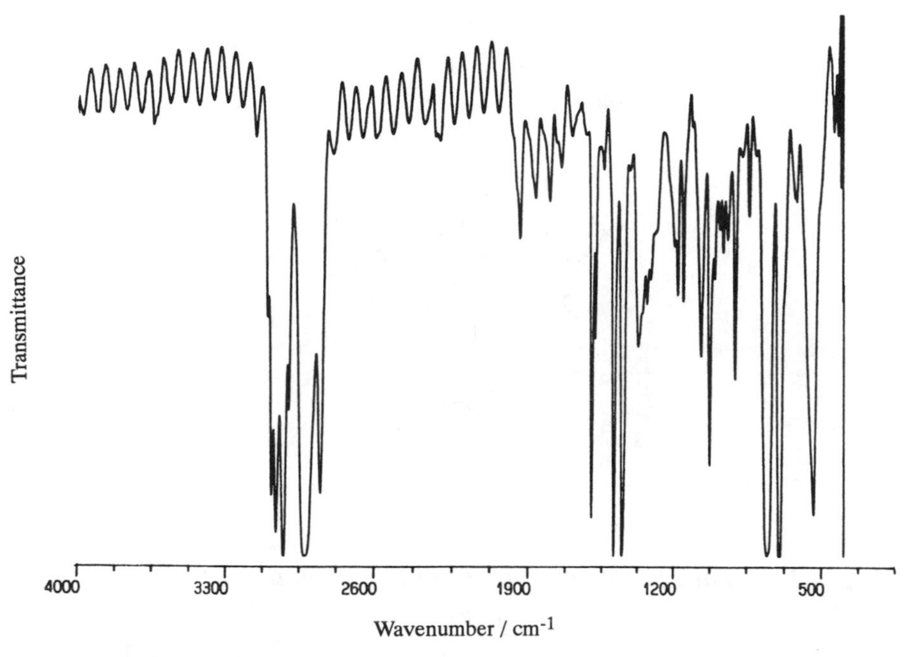

Fig. 2.2b. *A transmittance spectrum of polystyrene*

instruments fell away in the late 1960s when the improved technology of grating construction enabled cheap, good quality gratings to be manufactured.

2.3.1. Monochromators

The dispersive element is contained within a monochromator. Figure 2.3a shows the optical path of an infrared spectrometer which uses a grating monochromator.

Dispersion occurs when energy falling on the entrance slit is collimated on to the dispersive element. The dispersed element is then reflected back to the exit slit, beyond which lies the detector. The dispersed spectrum is scanned across the exit slit by rotating a suitable component within the monochromator. The width of the entrance and exit slits may be varied and programmed to compensate for any variation of the source energy with wavenumber. In the absence of a

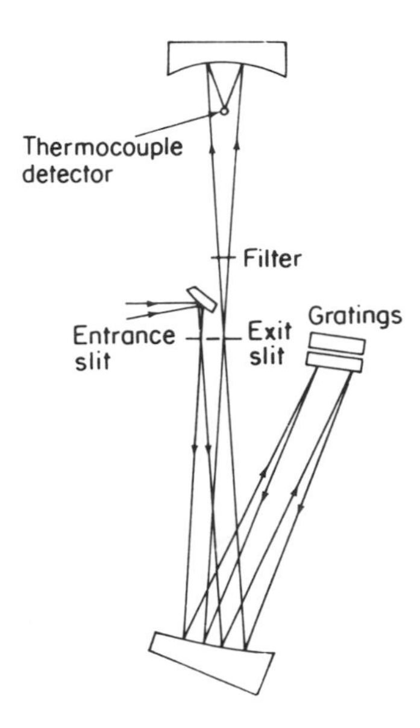

Fig. 2.3a. *The optical path of a double-beam infrared spectrometer with a grating monochromator*

sample the detector then receives radiation of approximately constant energy as the spectrum is scanned.

2.3.2. Double-beam Instruments

Atmospheric absorption by CO_2 and H_2O in the instrument beam has to be considered in the design of infrared instruments. Figure 2.3b shows the spectrum of these atmospheric absorptions. These contributions can be taken into account by using a double-beam arrangement in which radiation from a source is divided into two beams. These beams pass through a sample and a reference path of the sample compartment, respectively. The information from these beams is then ratioed to obtain the sample spectrum.

2.3.3. Sources and Detectors

A detector must have adequate sensitivity to the radiation arriving from the sample and monochromator over the entire spectral region

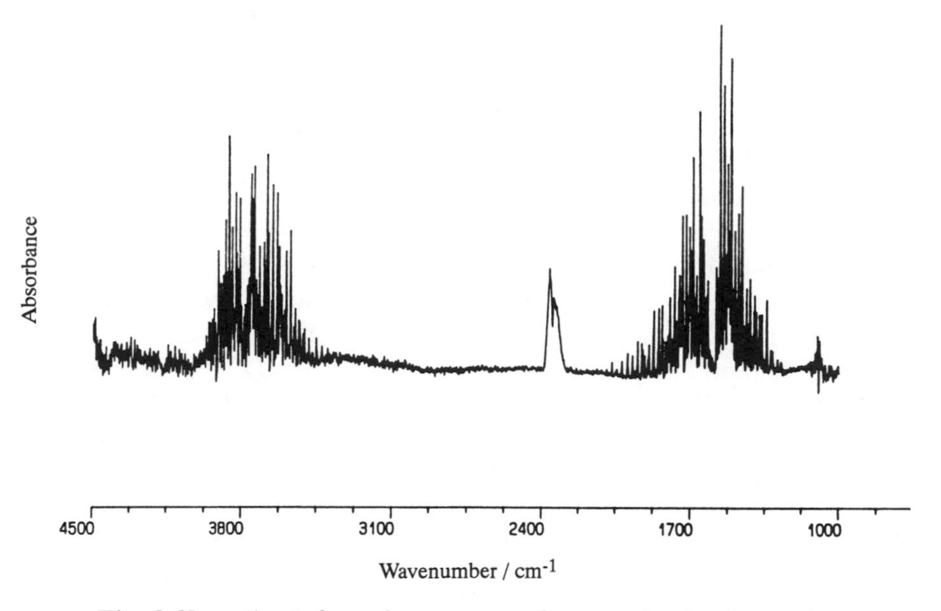

Fig. 2.3b. *An infrared spectrum of atmospheric absorptions*

required. In addition, the source must be sufficiently intense over both the wavenumber and transmittance range.

Sources of infrared emission have included the *Globar*, which is constructed of silicon carbide. There is also the *Nernst filament*, which is a mixture of the oxides of zirconium, yttrium and erbium. However, a Nernst filament only conducts electricity at elevated temperatures.

Most detectors have consisted of thermocouples with varying characteristics.

2.3.4. Limitations

The essential problem of the dispersive spectrometer lies with its monochromator. This contains narrow slits at the entrance and exit which limit the wavenumber range of the radiation reaching the detector to one resolution width. Samples from which a very quick measurement is needed, for example, in the eluant from a chromatography column, cannot be studied with instruments of low sensitivity because they cannot scan at speed. However, these limitations can be avoided through use of a Fourier-transform infrared spectrometer.

2.4. FOURIER-TRANSFORM INFRARED SPECTROMETERS

In recent years Fourier-transform infrared (FT-IR) spectroscopy has found increasing favour in laboratories. This more recent method is based on the old idea of the interference of radiation between two beams to yield an *interferogram*. An interferogram is a signal produced as a function of the change of pathlength between the two beams. The two domains of distance and frequency are interconvertible by the mathematical method of *Fourier transformation*.

Although the basic optical component of FT-IR instruments, the *Michelson interferometer*, has been known for almost a century, it was not until recent advances in computing that the technique could be successfully applied.

The basic components of an FT-IR spectrometer are shown schematically in Figure 2.4a. The radiation emerging from the source is passed to the sample through an interferometer before reaching a detector. Upon amplification of the signal, in which high-frequency contributions have been eliminated by a filter, the data are converted to a digital form by an analog-to-digital converter and then transferred to the computer for Fourier transformation to be carried out.

2.4.1. The Michelson Interferometer

The most common interferometer used is a Michelson interferometer, which consists of two perpendicularly plane mirrors, one of which can

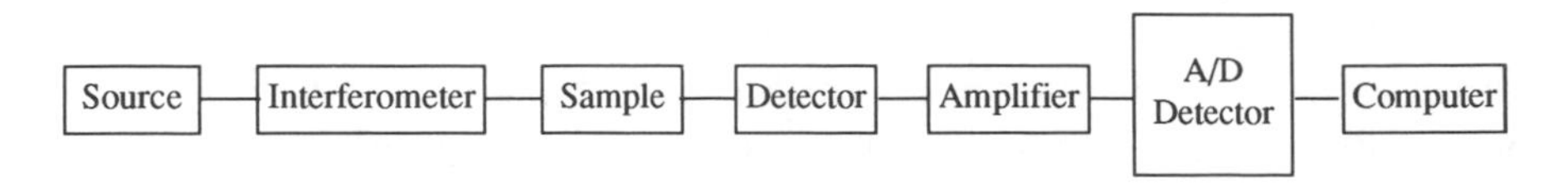

Fig. 2.4a. *The components of an FT-IR spectrometer*

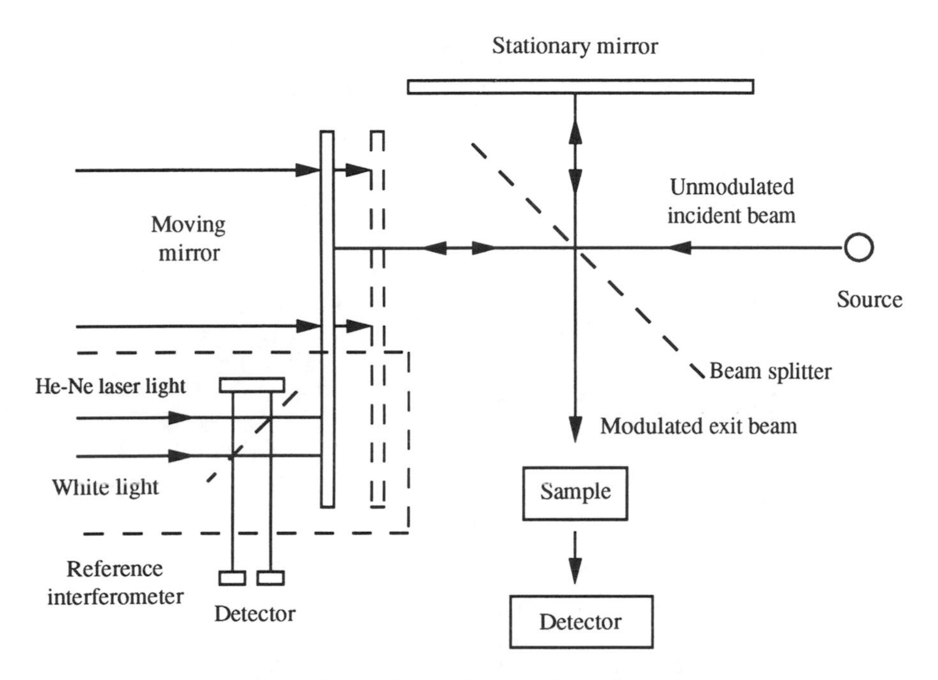

Fig. 2.4b. *A Michelson interferometer*

travel in a direction perpendicular to the plane (Figure 2.4b). A semi-reflecting film, the *beamsplitter*, bisects the planes of these two mirrors. The beamsplitter material has to be chosen according to the region to be examined. Materials such as germanium or iron oxide are coated on to an infrared-transparent substrate such as potassium bromide or caesium iodide to produce beamsplitters for the mid- or near-infrared regions. Thin organic films, such as poly(ethylene terephthalate), are used in the far-infrared region.

If a collimated beam of monochromatic radiation of wavelength λ is passed into an ideal beamsplitter, 50% of the incident radiation will be reflected to one of the mirrors and 50% will be transmitted to the other mirror. The two beams are reflected from these mirrors, returning to the beamsplitter where they recombine and interfere. Fifty per cent of the beam reflected from the fixed mirror is transmitted through the beamsplitter and 50% is reflected back in the direction of the source. The beam which emerges from the interferometer at 90° to the input beam is called the transmitted beam and this is the beam which is detected in FT-IR spectrometry.

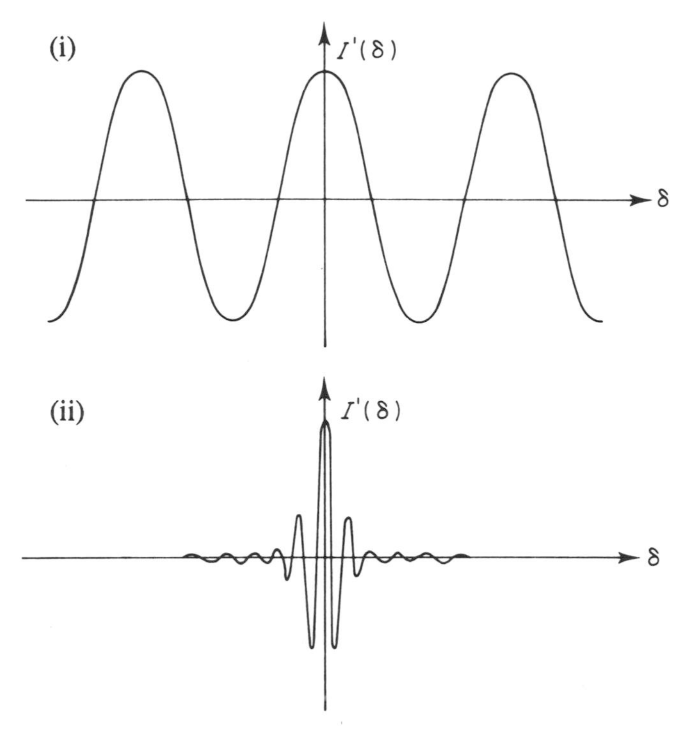

Fig. 2.4c. *Interferograms for (i) monochromatic radiation and (ii) polychromatic radiation*

The moving mirror produces an optical path difference between the two arms of the interferometer. For path differences of $(n + \frac{1}{2})\lambda$, the two beams interfere destructively in the case of the transmitted beam and constructively in the case of the reflected beam. The resultant interference pattern is shown in Figure 2.4c for a source of monochromatic radiation (i) and for a source of polychromatic radiation (ii). The former is a simple cosine function but the latter is of a more complicated form because it contains all the spectral information of the radiation falling on the director.

2.4.2. Fourier Transformation

The mathematics of the conversion of an interference pattern into a spectrum is beyond the scope of this book. However, we will look briefly at the process of Fourier transformation, without going into the detailed mathematics involved.

The essential equations relate the intensity falling on the detector, $I(\delta)$, to the spectral power density at a particular wavenumber $\bar{\nu}$, given by $B(\bar{\nu})$, and these are as follows:

$$I(\delta) = \int_{0}^{+\infty} B(\bar{\nu}) \cos 2\pi \bar{\nu} \delta \, d\bar{\nu} \tag{2.1}$$

which is one half of a cosine Fourier-transform pair, with the other being:

$$B(\bar{\nu}) = \int_{-\infty}^{+\infty} I(\delta) \cos 2\pi \bar{\nu} \delta \, d\delta \tag{2.2}$$

These two equations are interconvertible and are known as a Fourier-transform pair. The first shows the variation in power density as a function of difference in pathlength, which is an interference pattern. The second shows the variation in intensity as a function of wavenumber. Each can be converted into the other by the mathematical method of Fourier transformation. Do not worry if your if your calculus is not up to these equations. It is not necessary to have a detailed knowledge of the mathematics involved in order to carry out experiments using an FT-IR spectrometer!

The essential experiment to obtain an FT-IR spectrum is to produce interferograms, both with and without a sample in the beam, and then transform these interferograms into spectra of (a) the source with sample absorptions and (b) the source without sample absorptions. The ratio of the former and the latter corresponds to a double-beam dispersive spectrum.

The major advance towards routine use in the mid-infrared region came with a new mathematical method (or algorithm) devised by Cooley and Tukey in 1965 for fast Fourier transformation. This was combined with advances in microcomputers which enabled these calculations to be carried out rapidly on-line.

2.4.3. Advantages of Fourier-transform Infrared Spectroscopy

Fourier-transform infrared spectrometers have several significant advantages over dispersive instruments. Two of these are the Fellgett (or multiplex) advantage and the Jacquinot (or throughput) advantage. The *Fellgett advantage* is due to an improvement in the signal-to-noise ratio (SNR) per unit time , which is proportional to the square root of the number of resolution elements being monitored, and results from the large number of resolution elements being monitored simultaneously. In addition, because FT-IR spectrometry does not require the use of a slit or other restricting device, particularly at low resolution, the total source output can be passed through the sample continuously. This results in a substantial gain in energy at the detector, which translates to higher signals and improved SNRs. This is known as the *Jacquinot advantage*.

Another strength of FT-IR spectrometry is its *speed advantage*. The mirror has the ability to move short distances quite rapidly, and this, together with the SNR improvements due to the Fellgett and Jacquinot advantages, make it possible to obtain spectra on a millisecond timescale. In interferometry the factor that determines the precision of the position of an infrared band is the precision with which the scanning mirror position is known. By using a helium–neon laser as a reference, the mirror position is known with extremely high precision.

2.4.4. Sources and Detectors

As with dispersive instruments, FT-IR spectrometers use a Nernst or Globar source for the mid-infrared region. If the far-infrared region is to be examined, then a high-pressure mercury arc lamp can be used.

There are two commonly used detectors. The normal detector for routine use is a pyroelectric device incorporating deuterium tryglycine sulphate (DTGS) in a temperature-resistant alkali halide window. For more sensitive work, mercury cadmium telluride (MCT) can be used, but then the detector has to be cooled to liquid nitrogen temperatures.

2.4.5. The Moving Mirror

The moving mirror is the most crucial component of the interferometer. It has to be accurately aligned and must be capable of scanning two distances so that the path difference corresponds to a known value. A number of factors associated with the moving mirror need to be considered when calculating an infrared spectrum.

The interferogram is in the form of an analogue signal at the detector which then has to be digitised in order that Fourier transformation into a conventional spectrum can be carried out. There are two particular sources of error in transforming the digitised information on the interferogram into a spectrum.

First, the transformation carried out in practice involves an integration stage over a finite displacement, rather than over an infinite displacement. The mathematical process of Fourier transformation assumes infinite boundaries. The consequence of this necessary approximation is that the apparent lineshape of a spectral line may be as shown in Figure 2.4d, where the main band area has a series of negative and positive lobes (or pods) with diminishing amplitudes.

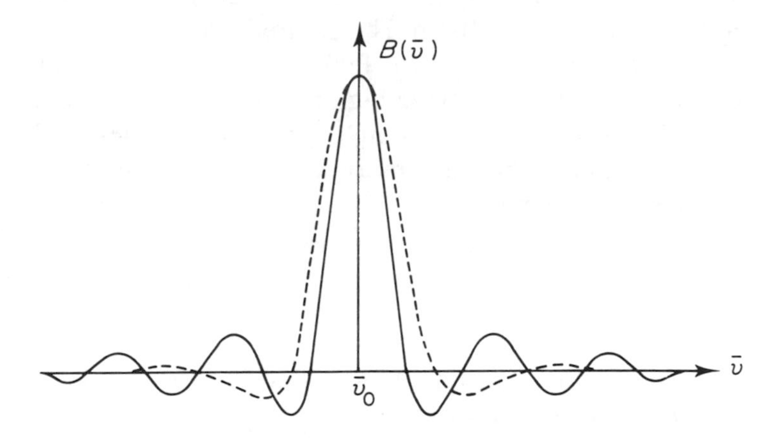

Fig. 2.4d. *Instrument lineshape without apodisation*

The process of *apodisation* is the removal of the side lobes or pods by multiplying the interferogram by a suitable function before the Fourier transformation is carried out. A suitable function must cause the intensity of the interferogram to fall smoothly to zero at both of its ends. Most FT-IR spectrometers offer a choice of apodisation options, with a good general purpose apodisation function being the following cosine function:

$$F(D) = \frac{1 + \cos(\pi D)}{2} \tag{2.3}$$

where D is the optical path difference. This cosine function provides a good compromise between the reduction in oscillations and the deterioration in spectral resolution. When accurate bandshapes are required, more sophisticated mathematical functions may be needed.

Another source of error arises if the sample intervals are not exactly the same on each side of the maxima corresponding to zero path difference. Phase correction is required, and this correction procedure ensures that the sample intervals are the same on each side of the first interval and should correspond to a path difference of zero.

The resolution is limited by the maximum path difference between the two beams. The limiting resolution in wavenumber (cm^{-1}) is the reciprocal of the pathlength difference (cm). For example, a pathlength difference of 10 cm is required to achieve a limiting resolution of 0.1 cm^{-1}. You may feel that this simple calculation indicates that it is easy to achieve high resolution. Unfortunately, this is not the case since the precision of both the optics and mirror movement mechanism become more difficult to achieve at longer displacements of pathlengths.

Summary

Various aspects of the instrumentation used in infrared spectroscopy were dealt with in this chapter.

Those aspects of modern dispersive infrared spectrometers which are necessary for understanding the principles of operation and the capabilities and limitations of typical instrumentation were discussed.

The operation and capabilities of Fourier-transform infrared spectrometers were also discussed. An understanding of the fundamental differences between dispersive and Fourier-transform instrumentation, including differences in performance, were examined.

Objectives

On completion of this chapter you should be able to:

- understand how an infrared spectrum is obtained from a dispersive instrument;
- understand how an infrared spectrum is obtained from a Fourier-transform instrument;
- appreciate the advantages and disadvantages of dispersive and Fourier-transform instruments;
- use an infrared spectrometer more effectively through a knowledge of the operating principles.

3. Sampling

3.1. INTRODUCTION

Infrared spectroscopy is a versatile analytical technique. It is relatively easy to obtain spectra from solids, liquids and gases. You have already studied the principles of infrared spectroscopy and the instrumentation required to produce the spectra. In this chapter we shall study how samples can be introduced into the instrument, the equipment required to obtain spectra and the pretreatment of samples.

First, we will examine the various ways of examining liquids, solids and gases using the traditional transmission methods of infrared spectroscopy. In the second part of the chapter we will examine the more modern reflectance methods which are in wide use today, such as the attenuated total reflectance, diffuse and specular reflectance methods. We will also look briefly at more specialist cells which you might encounter, such as photoacoustic, temperature and microsampling cells.

3.2. TRANSMISSION METHODS

Transmission spectroscopy is the oldest and most basic infrared method. The method is based upon the absorption of infrared radiation at specific wavelengths as it passes through a sample. It is possible to analyse samples in liquid, solid or gaseous form by using this approach.

3.2.1. Liquids and Solutions

There are several different types of transmission solution cells available. Fixed pathlength sealed cells are useful for volatile liquids,

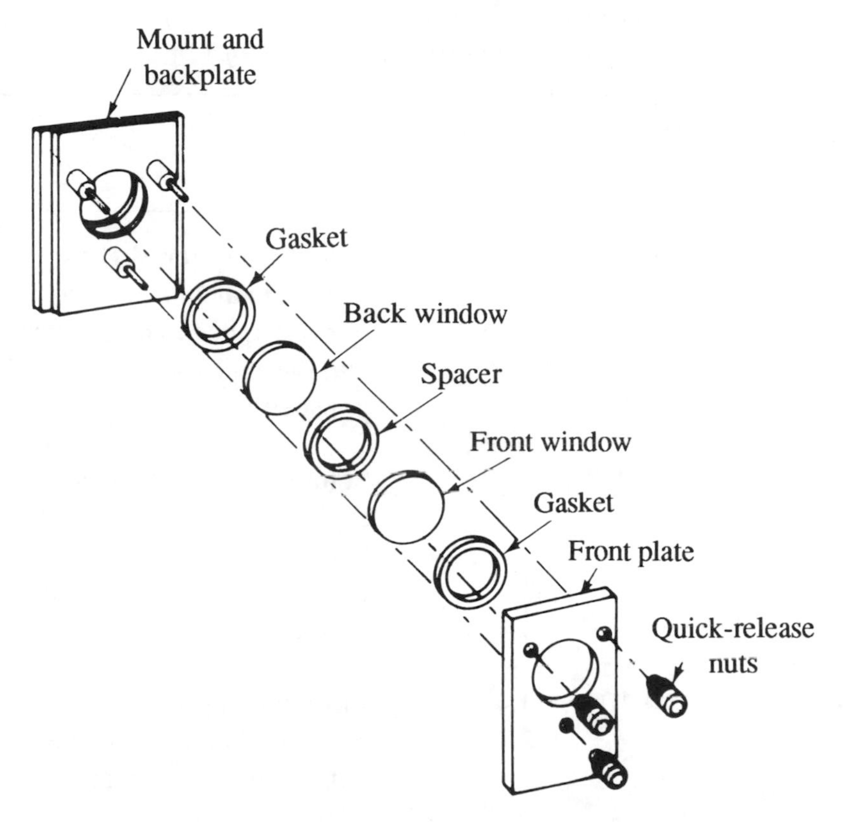

Fig. 3.2a. *A semi-permanent cell*

but cannot be taken apart for cleaning. Semi-permanent cells are demountable so that the windows can be cleaned. A semi-permanent cell is illustrated in Figure 3.2a. The spacer is usually made of polytetrafluoroethylene (PTFE) and is available in a variety of thicknesses, allowing one cell to be used for various pathlengths. Variable pathlength cells incorporate a mechanism for continuously adjusting the pathlength and a vernier scale allows accurate adjustment. All these cell types are filled by using a syringe.

An important consideration in the choice of infrared cells is the type of window material. The material must be transparent to the incident infrared radiation and therefore alkali halides are normally used in transmission methods. The cheapest material is sodium chloride (NaCl), with other commonly used materials being listed in Table

Table 3.2a. *Summary of the properties of some optical materials used in transmission infrared spectroscopy*

Window material	Useful range (cm^{-1})	Refractive index	Properties
NaCl	40 000–600	1.5	Soluble in water; slightly soluble in alcohol; low cost; fair resistance to mechanical and thermal shock; easily polished
KBr	43 500–400	1.5	Soluble in water and alcohol; slightly soluble in ether; hygroscopic; good resistance to mechanical and thermal shock
CaF_2	77 000–900	1.4	Insoluble in water; resists most acids and bases; does not fog; useful for high pressure work
BaF_2	66 666–800	1.5	Insoluble in water; soluble in acids and NH_4Cl; does not fog; sensitive to thermal and mechanical shock
KCl	33 000–400	1.5	Similar properties to NaCl but less soluble; hygroscopic
CsBr	42 000–250	1.7	Soluble in water and acids; hygroscopic
CsI	42 000–200	1.7	Soluble in water and alcohol; hygroscopic

3.2a.

SAQ 3.2a

What would be an appropriate material for liquid cell windows if you were wanting to examine an aqueous solution at pH 7?

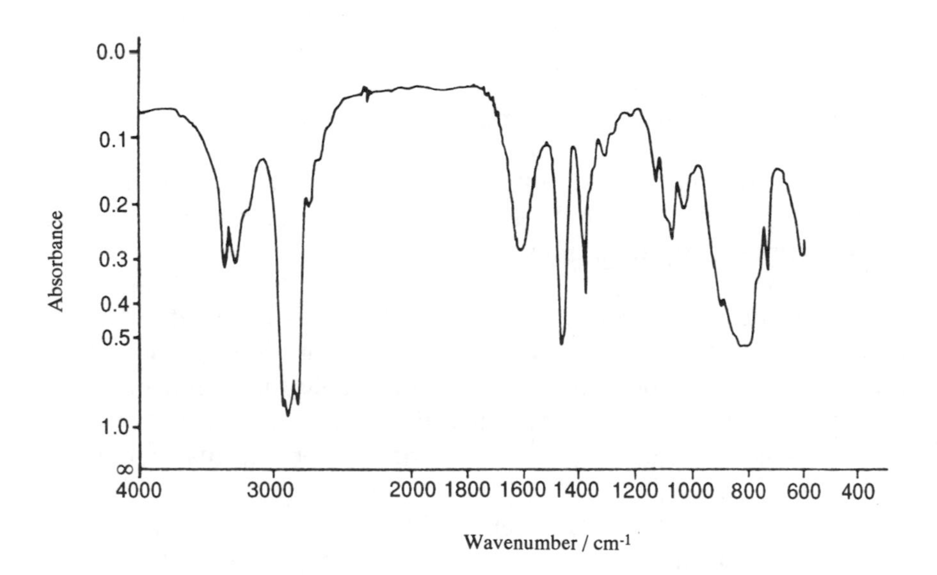

Fig. 3.2b. *An infrared spectrum of a hexylamine liquid film*

Liquid films provide a quick method for examining liquid samples. A drop of liquid may be sandwiched between two infrared plates, which are then mounted in a cell holder. The type of spectrum you would expect to produce from a liquid film is illustrated in Figure 3.2b. This shows the spectrum of hexylamine.

Π This method is not normally used for volatile (i.e. boiling point less than 100 °C) liquids. Why do you think this is?

A common problem encountered in obtaining good quality spectra from liquid films is sample volatility. When the spectrum of a volatile sample is recorded it progressively becomes weaker because evaporation takes place during the recording period. Liquids with boiling points below 100°C should be recorded in solution or in a short-pathlength sealed cell.

Before producing an infrared sample in solution, a suitable solvent must be chosen. In choosing a solvent for your sample you need to consider the following points:

(a) it has to dissolve the compound;

(b) it should be as non-polar as possible to minimise solute–solvent interactions;

(c) it should not strongly absorb infrared radiation.

If quantitative analysis is required it is necessary to use a cell of known pathlength. A guide to pathlength selection for different solution concentrations is shown in Table 3.2b.

In order to obtain the spectrum of a solution it is necessary to record spectra of both the solution and the solvent alone. The solvent spectrum may then be subtracted from the solution spectrum. There are two approaches — you may use the solvent alone in the cell as your reference in a double-beam experiment, or you may record the solvent spectrum separately and then subtract. Figure 3.2c, where the strong infrared spectrum of water has been removed from a relatively weak spectrum of a 1% w/v solution of aspirin, illustrates how subtraction can be very useful in FT-IR spectroscopy. You need to

Table 3.2b. *Pathlength selection for solution cells*

Concentration (%)	Pathlength (mm)
>10	0.05
1–10	0.1
0.1–1	0.2
<0.1	>0.5

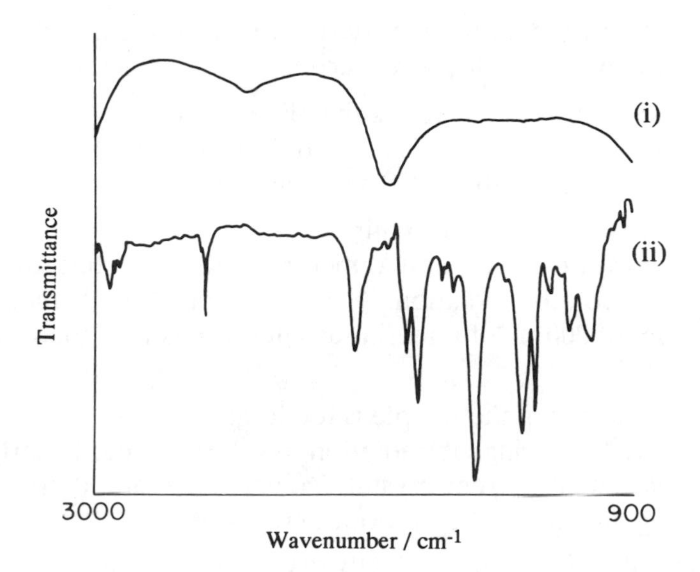

Fig. 3.2c. *The infrared spectra of (i) a 1% w/v solution of aspirin in water and (ii) the same solution after subtraction of the water spectrum*

take care, however, when subtracting spectra. It should be remembered that the concentration of the solvent alone is greater than that of the solvent in the solution and negative peaks may appear in the regions of solvent absorption.

3.2.2. Solids

There are three general methods of examining solid samples in the infrared: alkali halide discs, mulls and films. Your choice of method depends very much on the nature of the sample to be examined.

Alkali Halide Discs

This method involves mixing a solid sample (a few mg) with a dry alkali halide powder (100–200 mg). The mixture is usually ground with an agate mortar and pestle and subjected to a pressure of about 10 ton in^{-2} (1.575×10^5 kg cm^{-2}) in an evacuated die. This sinters the mixture and produces a clear transparent disc. The most commonly used alkali halide is potassium bromide (KBr), which is completely transparent in the mid-infrared region.

Certain factors need to be considered when preparing alkali halide discs. The following problems may arise:

(a) The ratio of the sample to alkali halide is wrong.
Surprisingly little sample is needed. You should use around 2 to 3 mg of sample with about 200 mg of halide.

(b) The disc is too thick or too thin.
Thin discs are fragile and difficult to handle, while thick discs transmit too little radiation. A disc of about 1 cm diameter made from about 200 mg of material usually results in a good thickness of about 1 mm.

(c) The crystal size of the sample is too large.
Excessive scattering of radiation results and particularly so at high wavenumbers. The crystal size must be reduced, normally by grinding the solid using a mortar and pestle.

(d) The alkali halide is not perfectly dry.
This results in the appearance of bands due to water. It is difficult to avoid, but the alkali halide should be kept desiccated and warm prior to use.

Mulls

This method for solid samples involves grinding the sample and then suspending it (about 50 mg) in 1–2 drops of a mulling agent. This is followed by further grinding until a smooth paste is obtained. Ideally, a mulling agent should be infrared transparent between 4000 and 600 cm^{-1}, but no such agent exists. The most commonly used mulling agent is Nujol (liquid paraffin), and its spectrum is shown in Figure 3.2d.

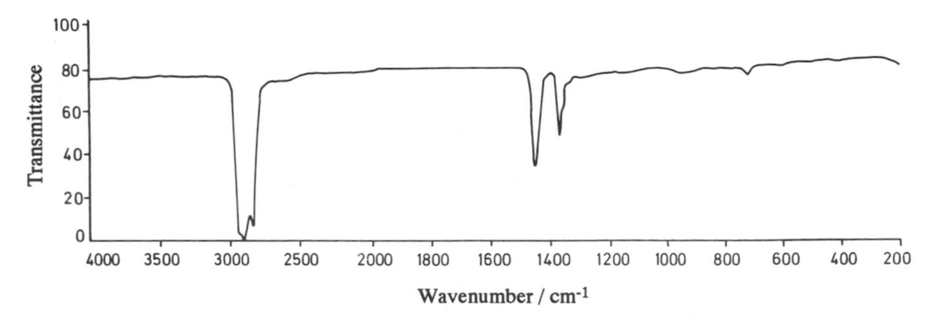

Fig. 3.2d. *The infrared spectrum of Nujol (liquid paraffin)*

Although the mull method is quick and easy, there are some potential problems:

(a) The ratio of the sample to mulling agent is wrong.
Too little sample and there will be no sign of the sample in the spectrum. Too much sample and a thick paste will be produced and no radiation will be transmitted. A rough guide to mull preparation is to use a microspatula tip of sample to 2–3 drops of mulling agent.

(b) The crystal size of the sample is too large.
This leads to a scattering of radiation which gets worse at the high-frequency end of the spectrum.

(c) The mull does not cover the whole beam.
If the mull is not spread over the whole plate area, part of the beam of radiation passes through the mull and consequently only part through the plate, thus producing a distorted spectrum.

(d) Too much or too little mull is placed between the infrared plates.
Too little mull leads to a very weak spectrum showing only the strongest absorption bands. Too much mull leads to poor transmission of radiation so that the baseline may be at 50% or less. It is sometimes possible to reduce the energy of a reference beam to a similar extent by use of an *attenuator*. Beam attenuators work somewhat like a venetian blind.

SAQ 3.2b

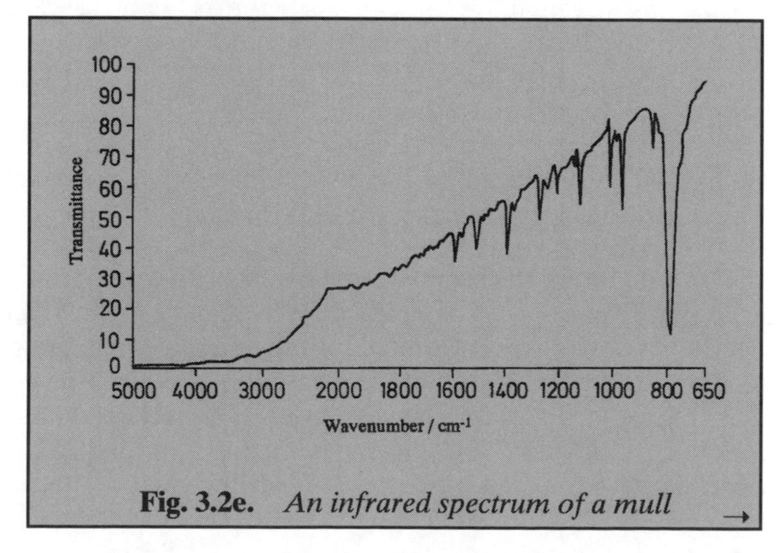

Fig. 3.2e. *An infrared spectrum of a mull*

→

SAQ 3.2b
(cont.)

Look at the spectrum of a mull in Figure 3.2e. What do you think is the problem with the mull produced and how would you go about remedying the problem?

Subtraction of spectra is especially useful for mulls, when the spectrum of the mulling agent can be subtracted, therefore giving the spectrum of the solid only. Figure 3.2f shows the spectrum of a Nujol mull of potassium benzoate, while the corresponding difference spectrum is shown in Figure 3.2g.

Films

Films can be produced by either solvent casting or by melt casting. These methods are particularly useful for examining polymers.

In solvent casting the sample is dissolved in an appropriate solvent, at a concentration which depends on the required film thickness. You need to choose a solvent which not only dissolves the sample, but will also produce a uniform film. The solution is poured on to a levelled glass plate (such as a microscope slide) or a metal plate and spread to a uniform thickness. The solvent is then evaporated in an oven and, once dry, the film can be stripped from the plate. Alternatively, it is possible to cast a film straight on to the infrared window to be used.

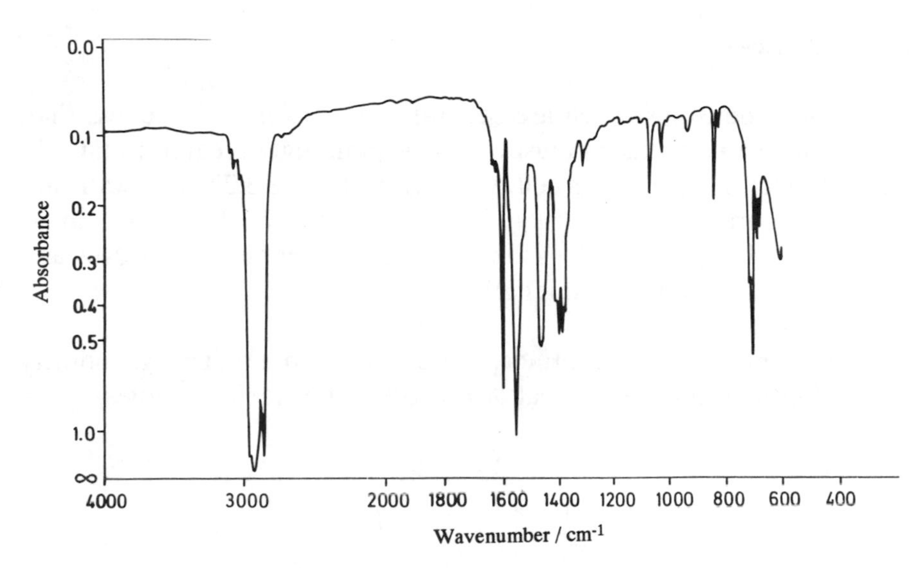

Fig. 3.2f. *The infrared spectrum of potassium benzoate in a Nujol mull*

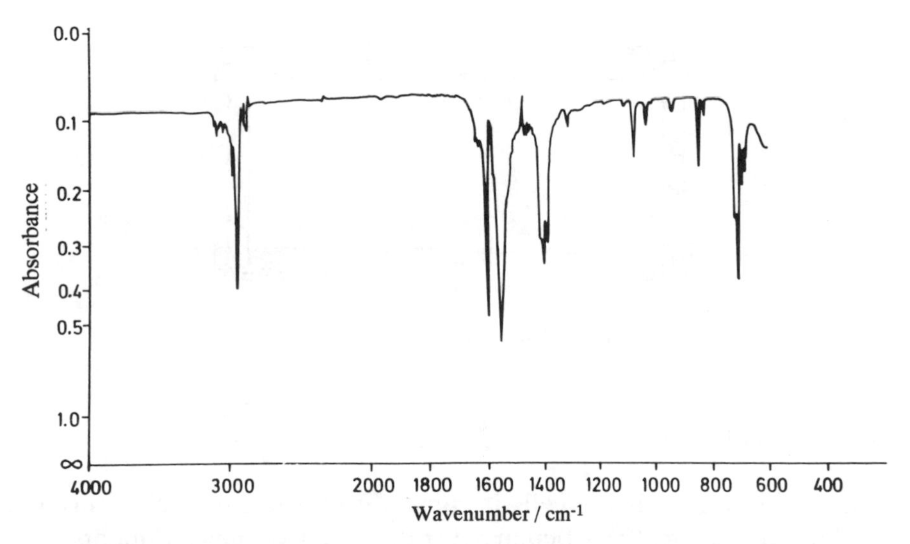

Fig. 3.2g. *The infrared spectrum of potassium benzoate after subtraction of the Nujol spectrum*

Solid samples which melt at relatively low temperatures without decomposition can be prepared by melt casting. A film is prepared by hot-pressing the sample in a hydraulic press between heated metal plates.

3.2.3. Gases

Gases have densities which are several orders of magnitude less than liquids, hence pathlengths must be correspondingly greater, usually 10 cm or longer. A typical gas cell is shown in Figure 3.2h. The walls are of glass or brass with the usual choice of windows. The cells can be filled by flushing or from a gas line. To analyse complex mixtures and trace impurities, longer pathlengths are necessary.

Typical applications are atmospheric pollution studies, gas purity determinations and free radical or reaction intermediate studies.

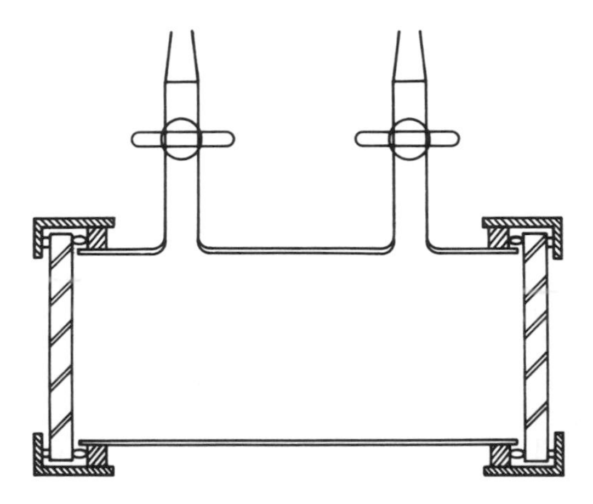

Fig. 3.2h. *A typical gas cell*

3.2.4. Pathlength Calibration

When using transmission cells it can be useful to know precisely the pathlength of the cell, particularly for quantitative measurements. The cell pathlength can be measured by the method of counting interference fringes. If an empty cell with parallel windows is placed in the spectrometer and a wavelength range scanned, an interference pattern similar to Figure 3.2i will be obtained.

The amplitude of the waveform will vary from 2 to 15%, depending on the state of the windows. The relationship between the pathlength

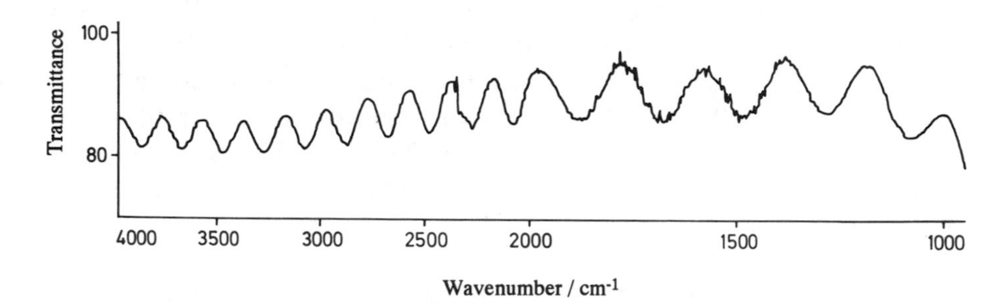

Fig. 3.2i. *An interference pattern recorded with an empty cell in the sample beam*

of the cell, L, and the peak to peak fringes is given by the following expression:

$$L = \frac{n}{2(\bar{\nu}_1 - \bar{\nu}_2)} \text{ cm} \tag{3.1}$$

where n is the number of complete peak to peak fringes between two maxima (or minima) at $\bar{\nu}_1$ and $\bar{\nu}_2$.

If the spectrometer is calibrated in wavelengths then the above equation can be converted into a more convenient form, as follows:

$$L = \frac{n\lambda_1\lambda_2}{2(\lambda_1 - \lambda_2)} \text{ cm} \tag{3.2}$$

where the values of the wavelength, λ, are expressed in cm.

When a beam of radiation is directed into the face of a cell, most of the radiation will pass straight through (beam A in Figure 3.2j). Some

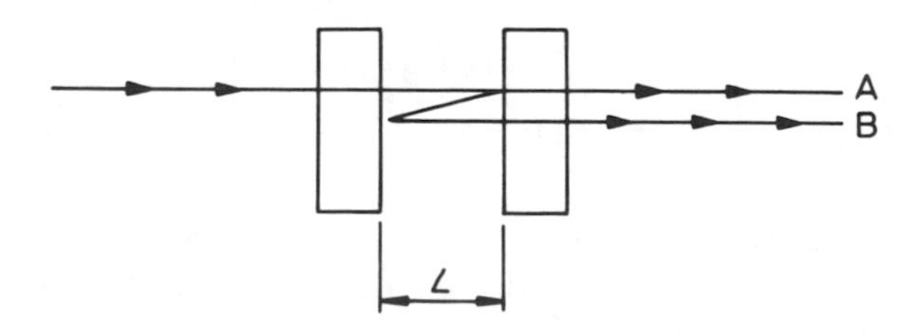

Fig. 3.2j. *A beam of radiation passing through an empty cell*

of the radiation will undergo a double reflection (beam B) and will therefore have travelled an extra distance $2L$ compared to beam A. If this extra distance is equal to a whole number of wavelengths then beams A and B will be in phase and the intensity of the transmitted beam (A + B) will be at a maximum. The intensity will be at a minimum when the two component beams are half a wavelength out of phase.

SAQ 3.2c

Using the interference pattern below, calculate the pathlength of the cell.

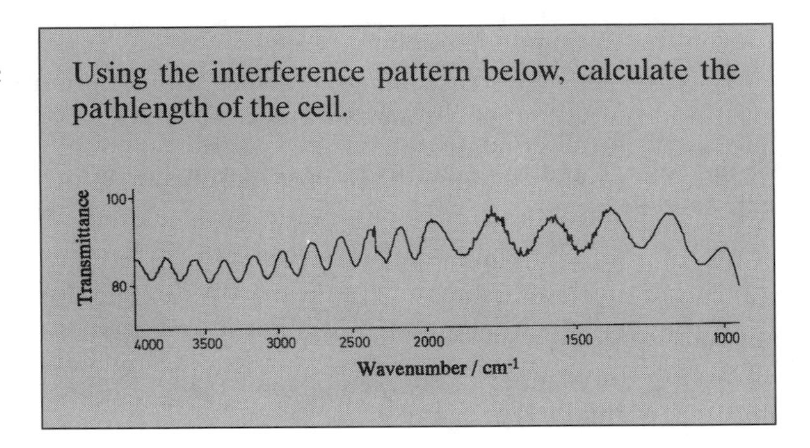

3.3. REFLECTANCE METHODS

Reflectance techniques can be used for samples which are difficult to analyse by the normal transmittance methods. Reflectance methods can be divided into two categories: internal reflectance measurements can be made by using an attenuated total reflectance cell in contact with the sample, while there are also external reflectance measurements which involve an infrared beam reflected directly from the sample surface.

3.3.1. Attenuated Total Reflectance Spectroscopy

Attenuated total reflectance (ATR) spectroscopy utilises the phenomenon of *total internal reflection* (Figure 3.3a). A beam of radiation entering a crystal will undergo total internal reflection when the angle of incidence at the interface between the sample and the crystal is greater than the critical angle. The *critical angle* is a function of the refractive indices of the two surfaces. The beam penetrates a fraction of a wavelength beyond the reflecting surface and when a material which selectively absorbs radiation is in close contact with the reflecting surface, the beam loses energy at the wavelength where the material absorbs. The resultant attenuated radiation is measured and plotted as a function of the wavelength by the spectrometer and gives rise to the absorption spectral characteristics of the sample.

The depth of penetration in ATR is a function of wavelength (λ), the refractive index of the crystal and the angle of incident radiation (θ). The depth of penetration, d_p, for a non-absorbing medium is given by the following formula:

$$d_p = \frac{\left(\lambda / n_1\right)}{2\pi \left[\sin\theta - \left(n_1 / n_2\right)^2\right]^{0.5}} \qquad (3.3)$$

where n_1 is the refractive index of the sample and n_2 is the refractive index of the ATR crystal.

The crystals used in ATR cells are made from materials which have low solubility in water and are of very high refractive index. Such materials include zinc selenide (ZnSe), germanium (Ge) and thallium/iodide (KRS-5). The properties of these commonly used materials for ATR crystals are summarised in Table 3.3a. Different designs of ATR cells allow both liquid and solid samples to

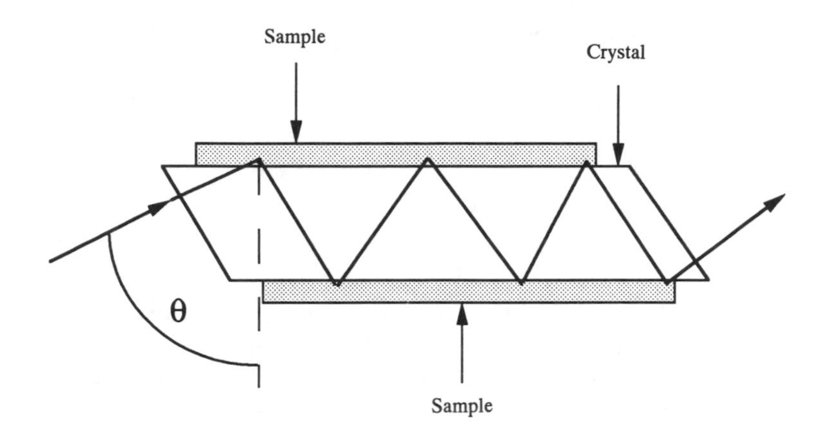

Fig. 3.3a. *An attenuated total reflectance cell*

be examined. The nature of the technique produces a less intense solvent contribution to the overall infrared spectrum and so solvent spectra can be easily subtracted from the sample spectrum of interest.

Table 3.3a. *Materials used as ATR crystals and their properties*

Window material	Useful range (cm^{-1})	Refractive index	Properties
KRS-5 (thallium/iodide)	17 000–250	2.4	Soluble in bases; slightly soluble in water; insoluble in acids; soft; highly toxic (handle with care)
ZnSe	20 000–500	2.4	Insoluble in water, organic solvents, dilute acids and bases
Ge	5000–550	4.0	Insoluble in water; very brittle

SAQ 3.3a

The spectrum of the polyamide film (refractive index 1.5) shown in Figure 3.3b was produced using an ATR cell made of KRS-5 (refractive index 2.4). If the incident radiation enters the cell crystal at an angle of 60°, what is the depth of penetration into the sample surface at:

(a) $1000\,cm^{-1}$;

(b) $1500\,cm^{-1}$;

(c) $3000\,cm^{-1}$?

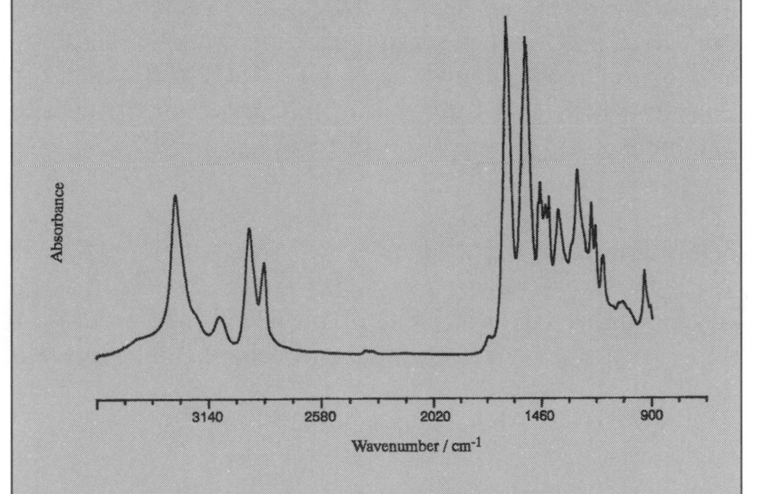

Fig. 3.3b. *An ATR spectrum of a polyamide film*

SAQ 3.3a

Multiple internal reflectance (MIR) and ATR are similar techniques, but MIR produces more intense spectra from multiple reflections. While a prism is usually used in ATR work, MIR uses specially shaped crystals which cause many internal reflections, typically 25 or more (Figure 3.3c).

3.3.2. Specular Reflectance

In external reflectance incident radiation is focused on to the sample and two forms of reflectance can occur, namely *specular* and *diffuse*.

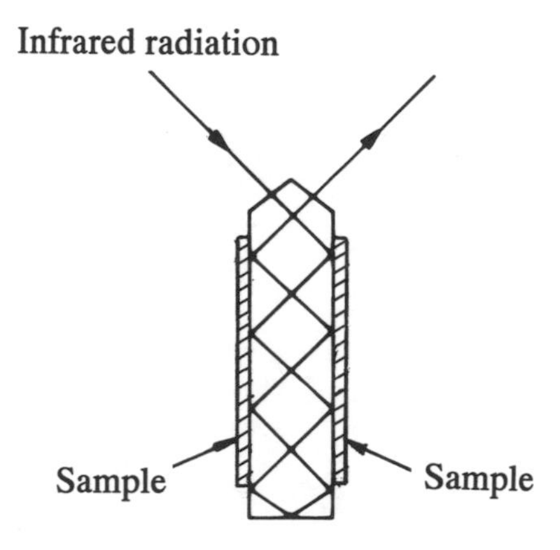

Fig. 3.3c. *A multiple internal reflectance cell*

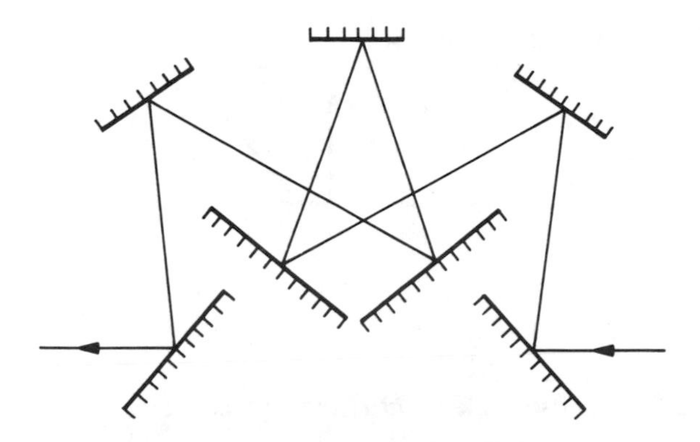

Fig. 3.3d. *Specular reflectance*

External reflectance measures the radiation reflected from a surface. The material must, therefore, be reflective or be attached to a reflective backing. A particularly useful application for this technique is the study of surface coatings such as surface treated metals, paints and polymers.

Specular reflectance occurs when the reflected angle of incident radiation equals the angle of incidence (Figure 3.3d). The amount of light reflected depends on the angle of incidence, the refractive index, the surface roughness and the absorption properties of the sample.

Increased pathlengths through thin coatings can be achieved by using grazing angles of incidence (up to 85°). This gives increased sensitivity. Thicker coatings in the micrometre thickness range are studied using angles of typically 30°.

3.3.3. Diffuse Reflectance

In external reflectance, the energy which penetrates one or more particles is reflected in all directions. This component is called *diffuse reflectance*. In the diffuse reflectance technique, commonly called DRIFT, a powdered sample is mixed with KBr powder. The DRIFT cell reflects radiation to the powder and collects the energy reflected back over a large angle. Diffusely scattered light can be collected directly from a sample or, alternatively, by using an abrasive sampling pad. DRIFT is most useful for sampling powders or fibres. Figure 3.3e

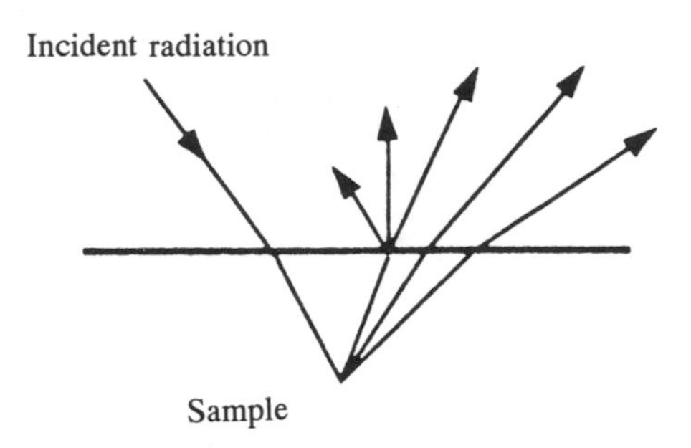

Fig. 3.3e. *Diffuse reflectance*

shows diffuse reflectance from the surface of a sample.

Kubelka and Munk developed a theory describing the diffuse reflectance process for powdered samples, which relates the sample concentration to the scattered radiation intensity. The Kubelka–Munk equation is as follows:

$$\frac{(1 - R_\infty)^2}{2R_\infty} = \frac{c}{k} \tag{3.4}$$

where R_∞ is the absolute reflectance of the layer, c is the concentration and k is the molar absorption coefficient.

Figure 3.3f shows the diffuse reflectance spectrum of a phenylene oxide–styrene copolymer. This copolymer is high melting, insoluble

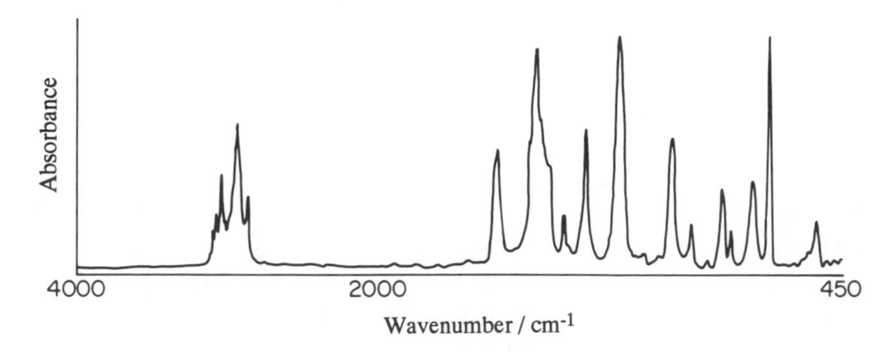

Fig. 3.3f. *A diffuse reflectance spectrum of a phenylene oxide–styrene copolymer*

and hard, but if rubbed with emery paper the powder deposited may be studied by diffuse reflectance.

3.4. OTHER TECHNIQUES

3.4.1. Photoacoustic Spectroscopy

Photoacoustic spectroscopy (PAS) is based on the transfer of modulated infrared radiation to a mechanical vibration. Gaseous, liquid or solid samples can be measured by using PAS and the technique is particularly useful for highly absorbing samples such as rubber or coal. When the modulated infrared radiation is absorbed by a sample, the substance heats and cools in response to the radiation reaching the sample. This heating and cooling pattern is converted into a pressure wave which can be detected by a microphone. Figure 3.4a shows a schematic diagram of a PAS cell.

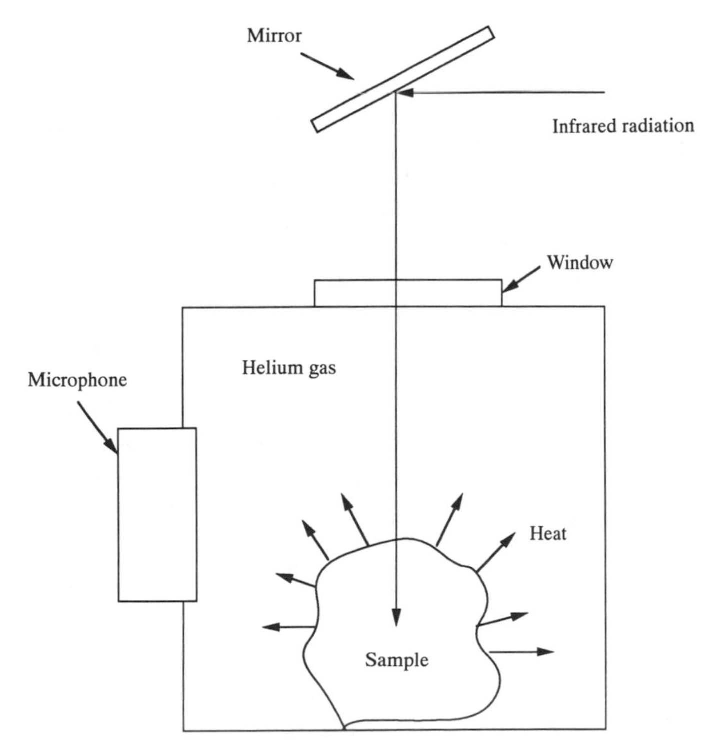

Fig. 3.4a. *Schematic representation of a photoacoustic spectroscopy cell*

Photoacoustic spectroscopy is useful because the detected signal is proportional to the sample concentration and can be used with very black or highly absorbing samples. The technique probes mainly material on the surface or several micrometres below the surface of the sample and so is very useful for surface studies.

3.4.2. The Use of Temperature Cells

Variable-temperature cells can be obtained which are controlled to 0.1°C in the range −180 to 250°C. An electrical heating system is used for temperatures above ambient, and liquid nitrogen with a heater for low temperatures. These cells can be used to study the phase transitions and kinetics of reactions. As well as transmission temperature cells, variable-temperature ATR cells and cells for microsampling are also available.

3.4.3. Microsampling

Special sampling accessories are available to allow examination of microgram or microlitre amounts. This is accomplished by using a *beam condenser* so that as much as possible of the beam passes through the sample. Microcells are available with volumes of around 4 μl and pathlengths up to 1 mm.

There are a number of microsampling cells available. A *diamond anvil cell* uses two diamonds to compress a sample to a thickness suitable for measurement and to increase the surface area. A multiple internal reflectance cell can also be used, as this technique can produce strong spectra. The ability to obtain spectra from trace amounts of materials is illustrated in Figure 3.4b, where the samples have been deposited on a micro-MIR crystal.

3.4.4. Combination Techniques

Infrared spectroscopy has been combined with various other analytical techniques. Gas chromatography–infrared spectroscopy

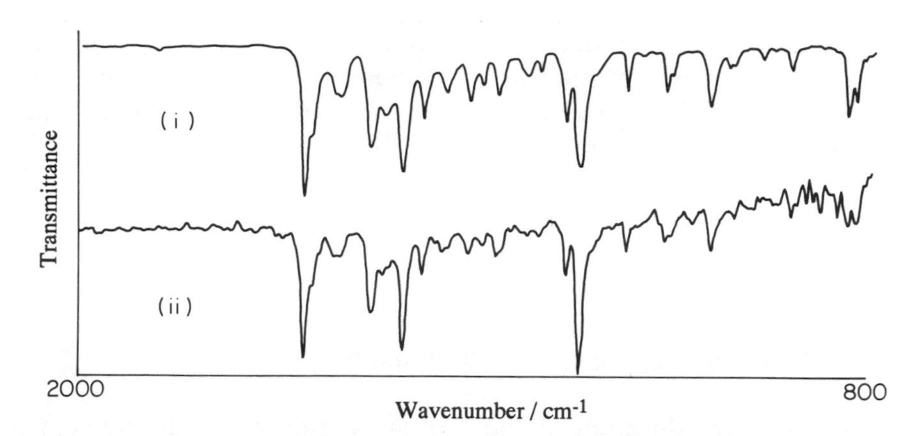

Fig. 3.4b. *The infrared spectra of phenacetin deposited on a micro multiple internal reflectance crystal: (i) 1 μg sample, spectrum recorded in 30 s; (ii) 50 ng sample, spectrum recorded in 2 min (resolution of 4 cm^{-1} in both cases)*

(GC–IR) allows the identification of the components eluting from a gas chromatograph, and has certain advantages over, for example gas chromatography–mass spectrometry (GC–MS). While the latter technique is able to distinguish easily between compounds of different mass, it is unable to differentiate structural isomers of the same molecular mass. By comparison, GC–IR can easily distinguish isomers. This technique is also particularly useful for pollution studies or gas purity determinations, for example.

Thermogravimetric analysis (TGA) is a technique which involves measuring the change of mass of a sample when it is heated. While TGA can provide quantitative information about a decomposition process, it is unable to identify the decomposition products. However, TGA and infrared spectroscopy have been combined to provide a complete qualitative and quantitative characterisation of thermal decomposition processes.

Summary

In this chapter transmission methods for obtaining infrared spectra were examined. The sampling methods which may be used for solids, liquids and gases were presented.

The various reflectance methods which are now widely available, such as ATR spectroscopy, specular reflectance and diffuse reflectance, were also introduced. Other techniques, such as PAS, GC–IR, and TGA–IR, plus the use of temperature cells and microsampling, were also described.

Objectives

On completion of this chapter you should be able to:

- recognise the different methods of sample preparation and sample handling techniques for solids, liquids and gases, which are used in infrared spectroscopy;
- understand the origins of reflectance techniques such as ATR spectroscopy, specular reflectance and diffuse reflectance;
- select appropriate sample preparation methods for different types of samples;
- recognise poor quality spectra and diagnose their causes;
- calculate the pathlength of cells from interference patterns.

4. Spectrum Interpretation

4.1. INTRODUCTION

Once you have recorded your infrared spectrum you have reached a crucial stage of this experimental technique, namely interpretation. Fortunately, spectrum interpretation is simplified by the fact that the bands which appear can be assigned to particular parts of the molecule, producing what are known as *group frequencies*. The mid-infrared region is normally treated as four regions, depending upon the type of group frequency. Each of these regions will be examined in turn in this chapter.

You should be aware, however, of certain factors which can complicate infrared spectra. Phenomena such as overtone and combination bands, Fermi resonance and hydrogen bonding can introduce additional, and sometimes misleading, information into the spectrum. It is important to be aware of these factors before tackling the interpretation of a given spectrum and so they are also discussed in this chapter.

4.2. GROUP FREQUENCIES

The mid-infrared spectrum can be divided into four regions and the nature of a group frequency may generally be determined by the region in which it is located.

4.2.1. The X—H Stretching Region (4000–2500 cm^{-1})

All fundamental vibrations in the region 4000–2500 cm^{-1} may be attributed to X—H stretching. For example, O—H stretching produces a broad band which occurs in the range 3700–3600 cm^{-1}. By comparison, N—H stretching is usually observed between 3400 and 3300 cm^{-1}. This absorption is generally much sharper than O—H stretching and may therefore be differentiated. Compounds containing the NH_2 group usually show a doublet structure, while secondary amines show one sharp band.

The C—H stretching bands from aliphatic compounds occur in the range 3000–2850 cm^{-1}. They are moderately broad and most organic compounds have many C—H bonds which result in medium intensity bands. It is also possible to resolve the asymmetrical and symmetrical C—H stretching absorptions of a CH_3 group, which usually occur at about 2965 and 2880 cm^{-1}, respectively, while the corresponding absorptions for a CH_2 group occur at 2930 and 2860 cm^{-1}.

Electronic effects of the neighbouring groups can affect the frequency of bands in this region. The C—H stretching frequency of the aldehyde group (H—C=O) is often split into two bands at 2850 and 2750 cm^{-1}. In addition, if the C—H bond is adjacent to a double bond or aromatic ring, the C—H stretching frequency increases and then absorbs between 3100 and 3000 cm^{-1}. This is useful in distinguishing purely aliphatic compounds, but must be used with care, as it is found that for many compounds containing a small number of aromatic hydrogens and many aliphatic C—H bonds the peaks above 3000 cm^{-1} may only appear as a shoulder on the stronger aliphatic absorption and may be obscured. Hydrogens attached to carbons in small strained rings (cyclopropane and cyclobutane) and to carbons carrying chlorine also absorb above 3000 cm^{-1} and this may lead to confusion. However, evidence for the presence of an aromatic ring may be obtained by examination of other regions of the spectrum. Hydrogen attached to an sp hybridised carbon gives rise to higher frequencies and the bands are easily interpreted. For example, C≡C—H stretching absorbs as a sharp, medium intensity, single peak at about 3300 cm^{-1}.

Deuterated compounds are expected to show C—D stretching at a frequency less than that observed for C—H stretching, due to the change in atomic mass (remember SAQ 1.3b). C—D stretching occurs at a factor of 0.73 less than that of C—H stretching, i.e. at about 2130 cm^{-1}.

SAQ 4.2a

Examine the spectra in Figures 4.2a–4.2e and classify them as below:

(i) aliphatic C—H bonds only;

(ii) aliphatic and aromatic C—H bonds;

(iii) an alkene or aromatic compound containing no aliphatic C—H bonds;

(iv) an alkyne;

(v) a deuterated compound.

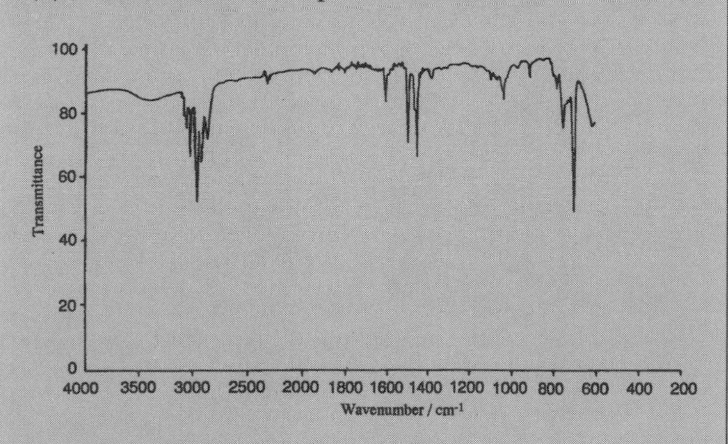

Fig. 4.2a. *Infrared spectrum of an unknown substance*

→

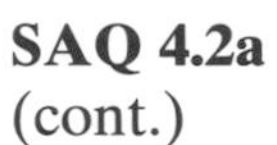

SAQ 4.2a
(cont.)

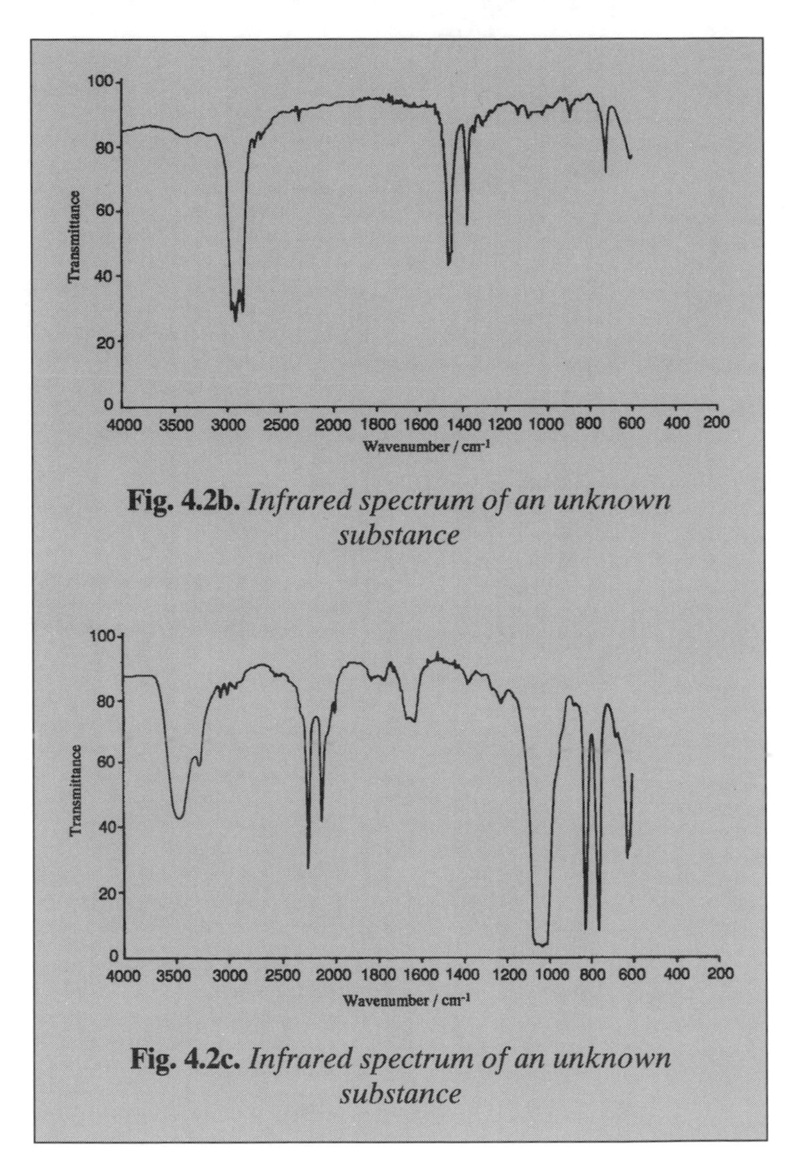

Fig. 4.2b. *Infrared spectrum of an unknown substance*

Fig. 4.2c. *Infrared spectrum of an unknown substance*

SAQ 4.2a
(cont.)

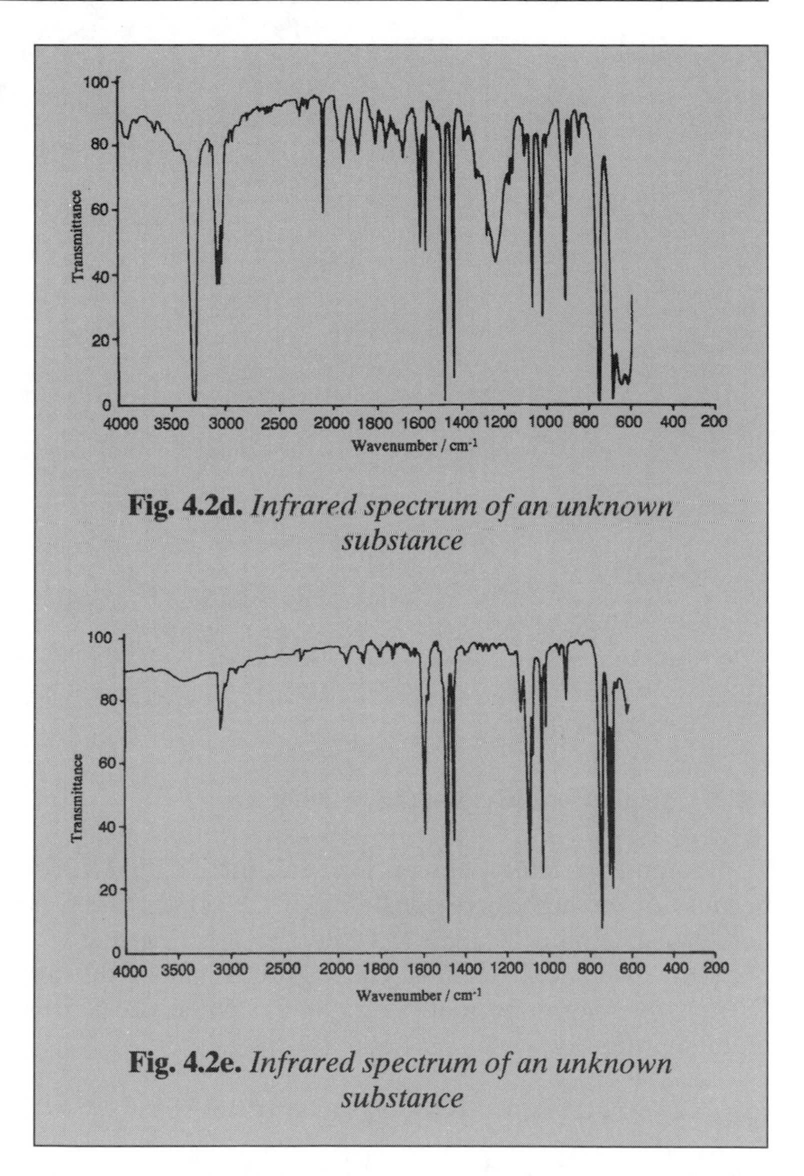

Fig. 4.2d. *Infrared spectrum of an unknown substance*

Fig. 4.2e. *Infrared spectrum of an unknown substance*

SAQ 4.2a

4.2.2. Triple Bond Region (2500–2000 cm^{-1})

Triple bond stretching absorptions fall in the 2500–2000 cm^{-1} region because of the high force constants of the bonds; C≡C bonds absorb between 2300 and 2050 cm^{-1}, while the nitrile group (C≡N) occurs between 2300 and 2200 cm^{-1}. These groups may be distinguished since C≡C stretching is normally very weak, while C≡N stretching is of medium intensity.

Π Can you think of a reason as to why this is so?

The change in dipole moment during absorption in a C≡C bond is likely to be very small, unless a polar group is attached. The C≡N group has a large dipole and hence there is a large change as the bond length is varied. The intensity of the absorption depends on this difference in dipole moment.

These are the most common absorptions in this region, but you may come across some X—H stretching absorptions where X is a more massive atom, such as phosphorus or silicon. These absorptions usually occur near 2400 and 2200 cm^{-1}, respectively.

SAQ 4.2b

> Carbon monoxide absorbs at 2143 cm^{-1}. What does this tell you about the bond order in this molecule?

4.2.3. Double Bond Region (2000–1500 cm^{-1})

The principal bands in this region are due to C=C and C=O stretching. Carbonyl stretching is one of the easiest absorptions to recognise in an infrared spectrum. It is usually the most intense band in the spectrum, and depending on the type of C=O bond present, occurs in the region 1830–1650 cm^{-1}. It should be noted, however, that metal carbonyls may absorb above 2000 cm^{-1}.

The C=C stretching bond is much weaker and occurs at around 1650 cm^{-1}. This band is often absent for symmetry or dipole moment reasons. The C=N stretching band also occurs in this region and is usually stronger.

The N—H bending vibration in amines occurs between 1630 and 1500 cm^{-1} and is usually strong. Before assigning a band always check the N—H stretching region above 3000 cm^{-1} to avoid any possible confusion.

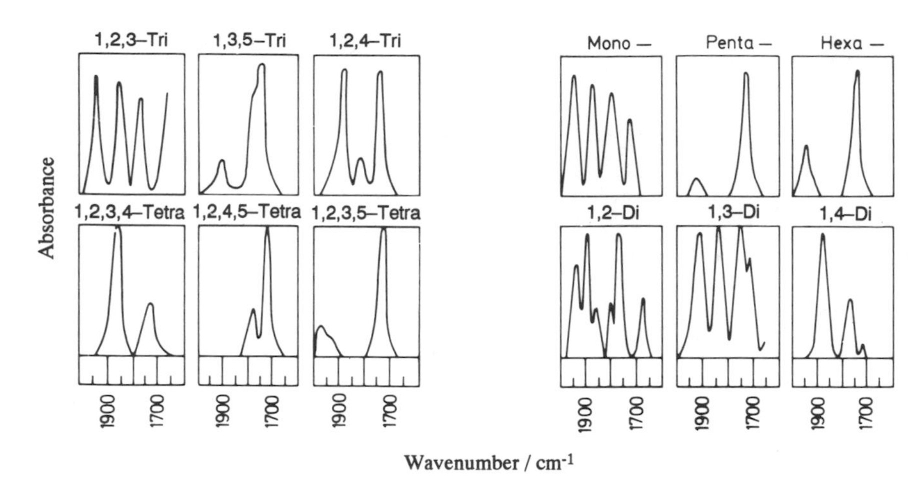

Fig. 4.2f. *C—H out-of-plane bending overtones and combination band patterns for substituted benzenes*

A series of weak bands in the region from 2000 to 1650 cm^{-1} are often used for assignment purposes. These are combination bands from substituted benzenes. Combination bands will be discussed in more detail in Section 4.3.1. It has been found that the intensity, frequency and number of absorptions in this region are a reliable index to the substitution pattern in the benzene ring. These patterns are illustrated in Figure 4.2f.

SAQ 4.2c

> The spectra in Figures 4.2g–4.2i are of 1,2-dimethylbenzene, 1,3-dimethylbenzene and 1,4-dimethylbenzene, in the region 2000–1650 cm^{-1}. Which is which?

SAQ 4.2c
(cont.)

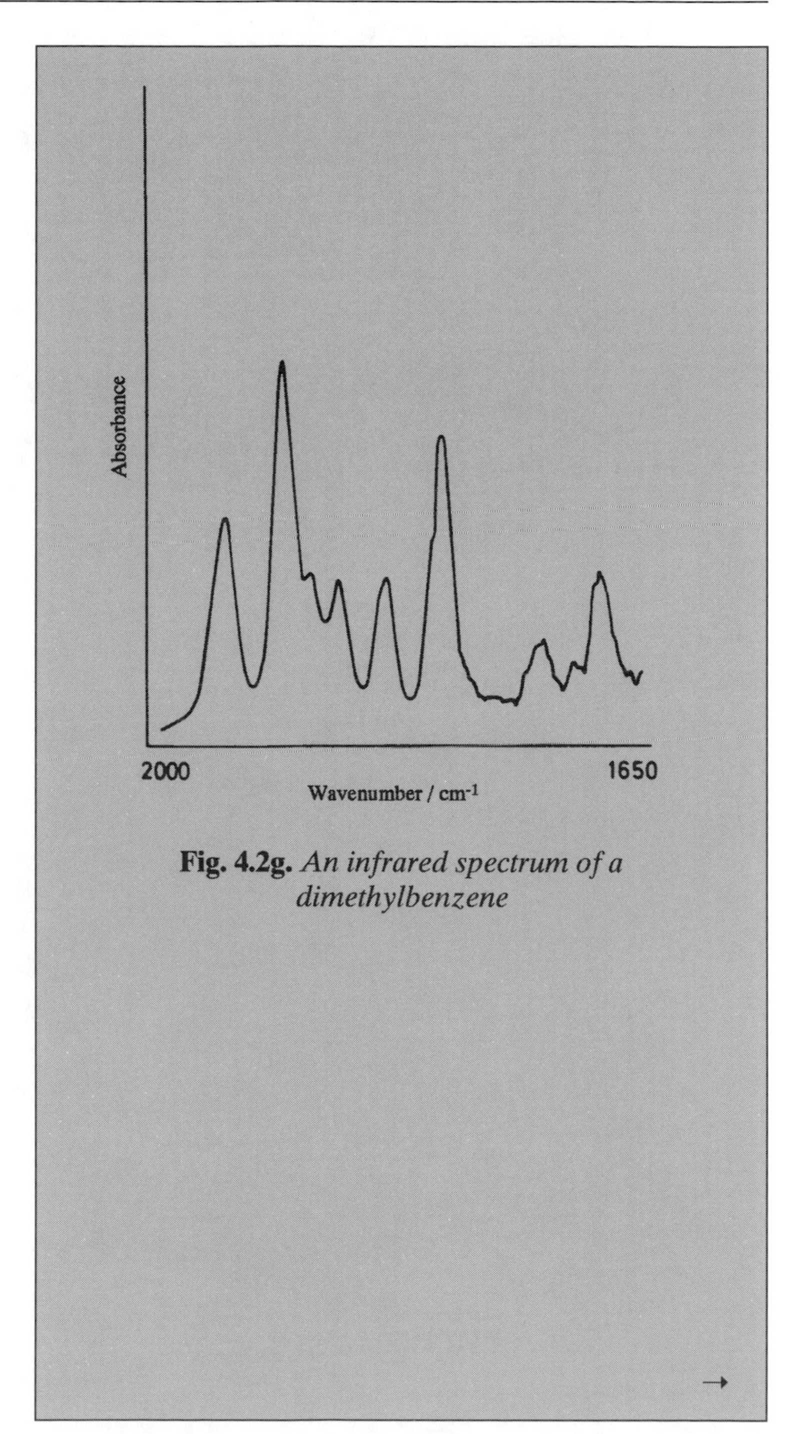

Fig. 4.2g. *An infrared spectrum of a dimethylbenzene*

→

SAQ 4.2c
(cont.)

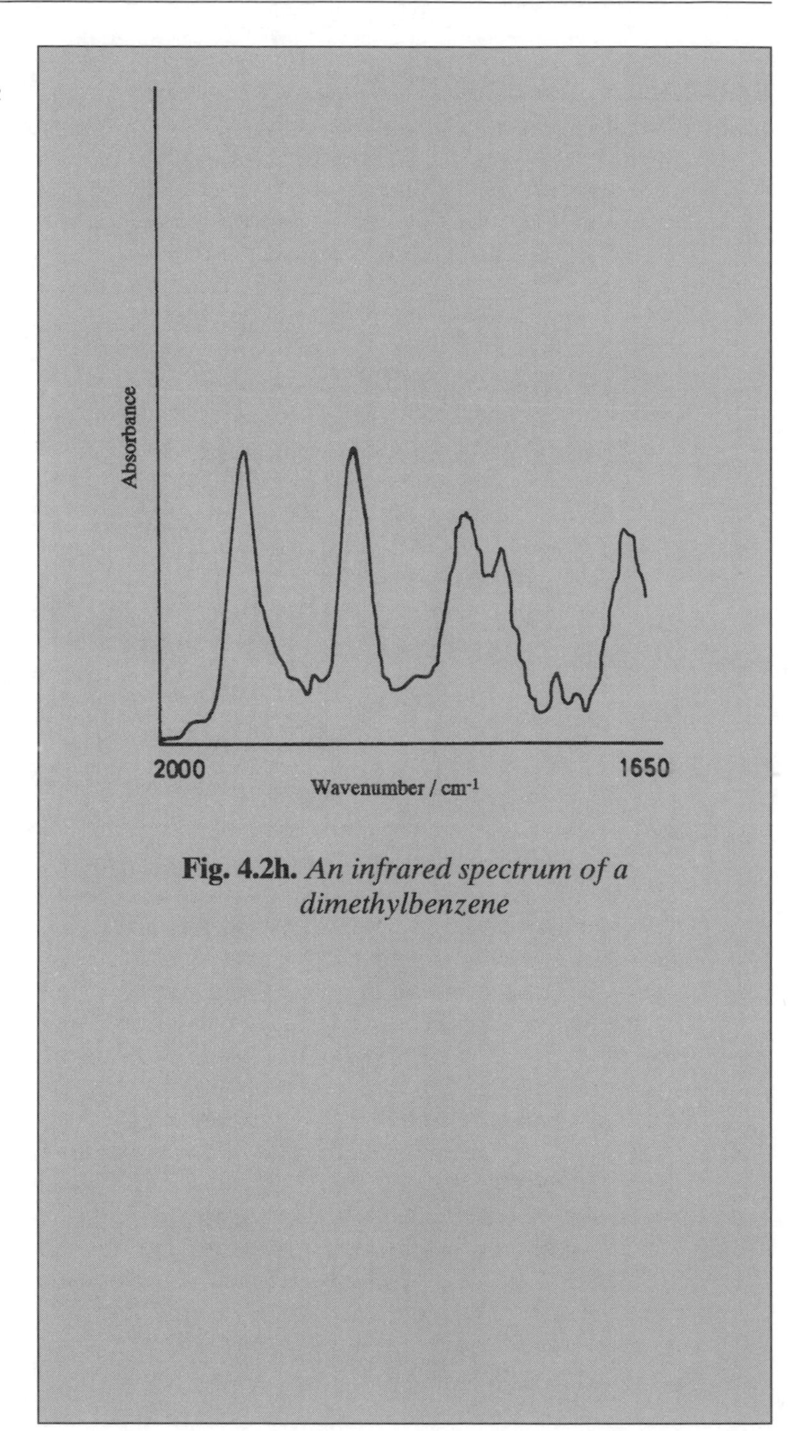

Fig. 4.2h. *An infrared spectrum of a dimethylbenzene*

SAQ 4.2c
(cont.)

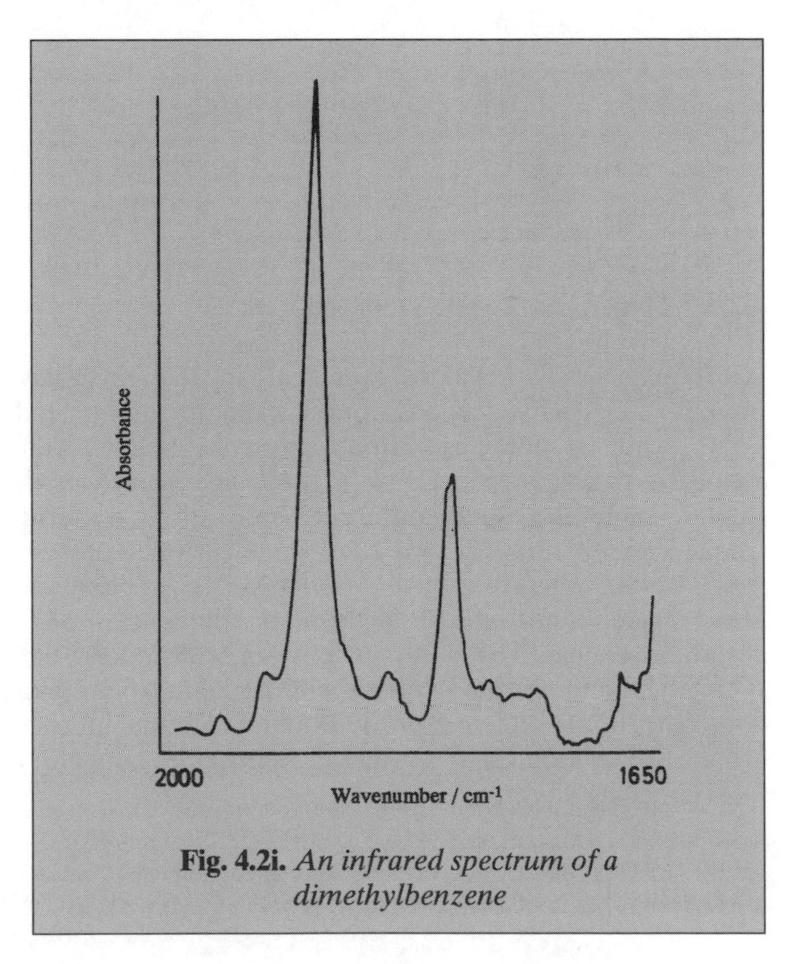

Fig. 4.2i. *An infrared spectrum of a dimethylbenzene*

SAQ 4.2c

4.2.4. Fingerprint Region (1500–600 cm^{-1})

At frequencies with values greater than 1500 cm^{-1} it is generally possible to assign each absorption band in an infrared spectrum. This is not so for absorptions observed below 1500 cm^{-1}. This particular region is referred to as the *fingerprint region*, since quite similar molecules give different absorption patterns at these frequencies.

Most single bonds absorb at similar frequencies and hence the vibrations couple. The observed pattern will depend on the carbon skeleton, and the resulting bands will originate from the oscillation of large parts of the skeleton, or the skeleton and the attached functional groups. In addition C—C stretching frequencies may also couple with C—H bending vibrations.

SAQ 4.2d

> Would you expect a C—O stretching mode to be more or less intense than a C—C stretching mode?

The C—O stretching frequency is one of the bands that can be particularly useful for identification purposes. If no intense band appears in the fingerprint region, you can usually be sure that no C—O bonds are present. The frequency is rather variable, occurring anywhere between 1400 and 1000 cm^{-1}.

Aromatic rings give rise to two bands at 1600 and 1500 cm^{-1}. They are usually sharp, but are of variable intensity, and occasionally the band at 1600 cm^{-1} splits into a doublet.

Aromatic rings and alkenes give rise to other bands which are perhaps the most useful in this region. These are out-of-plane C—H bending vibrations which occur between 1000 and 700 cm^{-1}. In alkenes the pattern varies, depending on the substitution pattern, so that *cis*- and *trans*- alkenes containing the $=CH_2$ group may be differentiated. In substituted benzenes the spectral pattern gives information about the substitution pattern in the ring, since bands characteristic of one, two or three adjacent C—H bonds appear. This means that 1,2-substitution gives a different pattern from 1,3- and 1,4-substituted ring systems. Note, however, that C—Cl stretching occurs at around 700 cm^{-1} and is easily confused with C—H out-of-plane bending for aromatic rings.

The nitro group (NO_2) gives two strong peaks at 1475 and 1550 cm^{-1}.

The information provided by these four regions can be summarised in what are known as correlation tables. Some useful correlation tables appear in the Appendix.

4.3. COMPLICATING FACTORS

Although it is a useful assumption to make, it is not true that all bands in a spectrum can simply by associated with a particular bond or part of a molecule. Other complicating factors have to be taken into account.

4.3.1. Overtone and Combination Bands

The sound we hear is a mixture of harmonics — a fundamental frequency mixed with multiples of that frequency. *Overtone bands* in

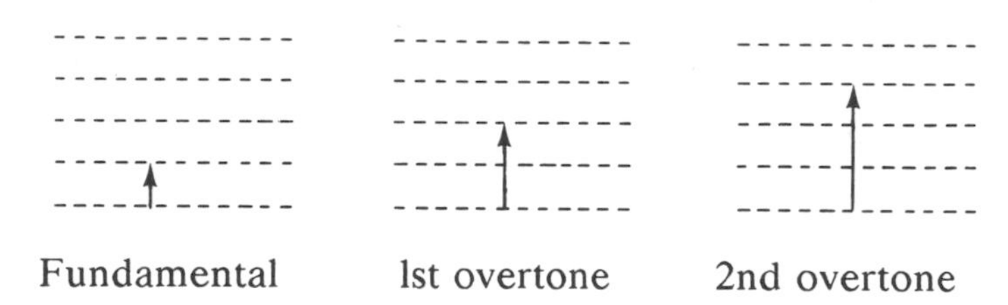

Fig. 4.3a. *Energy levels for fundamental and overtone infrared bands*

an infrared spectrum are analogous — they are multiples of the fundamental absorption frequency.

The energy levels for overtones of infrared modes are illustrated in Figure 4.3a.

The energy required for the first overtone is twice the fundamental, assuming evenly spaced energy levels. Since the energy is proportional to the frequency absorbed and this is proportional to the wavenumber, the first overtone will appear in the spectrum at twice the wavenumber of the fundamental.

Combination bands arise when two fundamental bands absorbing a $\nu_1 + \nu_2$ absorb energy simultaneously. The resulting band will appear at $(\nu_1 + \nu_2)$ wavenumbers.

SAQ 4.3a

A molecule has strong fundamental bands at the following frequencies:

C—H bending at 730 cm^{-1};
C—C stretching at 1400 cm^{-1};
C—H stretching at 2950 cm^{-1}.

Write down the frequencies of the possible combination bands and the first overtones.

SAQ 4.3a

4.3.2. Fermi Resonance

The *Fermi resonance* effect usually leads to two bands appearing close together when only one is expected. When an overtone or a combination band has the same or similar frequency as a fundamental, two bands appear, split either side of the expected value, and are of about equal intensity. The effect is greatest when the frequencies match, but it is also present when there is a mismatch of a few tens of wavenumbers. The two bands are referred to as a Fermi doublet.

SAQ 4.3b

Tetrachloroethane is expected to show only four infrared-active fundamentals. Three of these fundamentals absorb at 217, 313 and 459 cm^{-1}. The fourth is expected to occur in the region 700–800 cm^{-1}. The spectrum has two bands in this frequency range at 762 and 791 cm^{-1}. Can you account for this observation?

SAQ 4.3b

4.3.3. Hydrogen Bonding

Hydrogen bonding can be defined as the attraction which occurs between a highly electronegative atom carrying a non-bonded electron pair (such as fluorine, oxygen or nitrogen) and a hydrogen atom, itself bonded to a small highly electronegative atom. An example of this type of bonding is illustrated by the interactions between water molecules (Figure 4.3b).

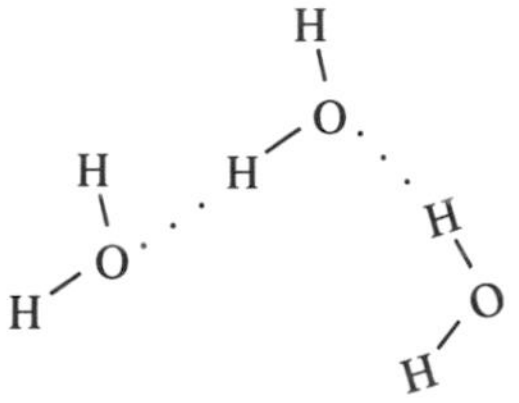

Fig. 4.3b. *Hydrogen bonding of water molecules*

This is an example of *intermolecular* hydrogen bonding. It is also possible for a hydrogen bond to form between the OH and COOH groups within the same molecule. This is known as *intramolecular* hydrogen bonding and is illustrated by the structural formulae shown in Figure 4.3c.

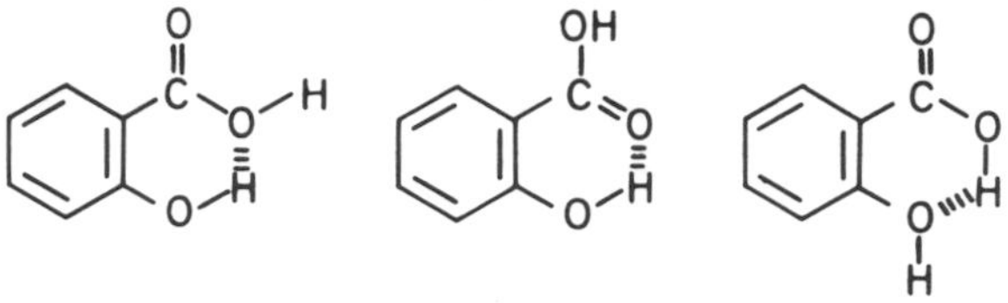

Fig. 4.3c. *Intramolecular hydrogen bonding*

Many solvents are capable of forming hydrogen bonds to solutes. To form a hydrogen bond we need a proton donor group and an electron donor group. Three classes of solvent exist which may lead to trouble when used for hydrogen bonding studies is solution:

1. Compounds which contain hydrogen donor groups, e.g. halogenated compounds which contain a sufficient number of halogens to activate the hydrogens present, such as chloroform.

2. Compounds which contain non-bonded electron pairs. In this class are ethers, aldehydes and tertiary amines.

3. Compounds which contain both types of group, e.g. water and alcohols.

There are only a few solvents that do not have the above characteristics, such as carbon tetrachloride (CCl_4) and carbon disulphide (CS_2). These still contain lone electron pairs but being on S and Cl they are less available, and any interaction will therefore be extremely weak.

Hydrogen bonding is a very important effect in infrared spectroscopy. We know that in infrared spectroscopy the frequencies of vibration of bonds depend on the masses of the atoms in the bond and the bond stiffness. Hydrogen bonding influences the bond stiffness and so alters the frequency of vibration. For example, if we look at a hydrogen bond in an alcohol, the O—H stretching vibration in a hydrogen-bonded dimer is observed in the range 3500–2500 cm^{-1}, rather than over the usual range 3700–3600 cm^{-1}.

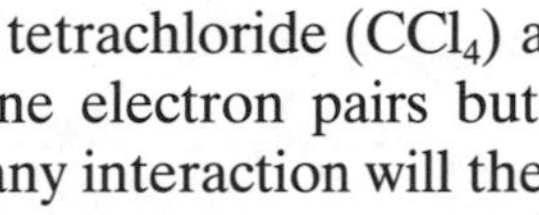

R–O—H•••O O–H stretching (3500–2500 cm^{-1})

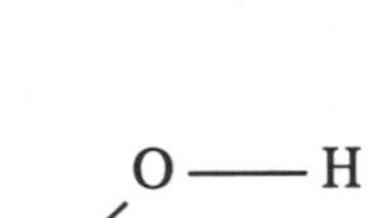

R–O—H O–H stretching (3700–3600 cm^{-1})

Apart from solvent effects, concentration and temperature also affect the degree of hydrogen bonding in a compound. The lower the concentration, then the less chance there is of two molecules colliding. It follows that the degree of hydrogen bonding decreases with decreasing concentration.

Π What do you think would be the effect of an increase in temperature on the infrared spectrum of a hydrogen-bonded compound?

Increasing temperature means that each molecule will have more energy on average and hence weak associative forces are likely to be broken. This should lead to a lesser degree of hydrogen bonding, and thus changes in frequency to greater values would be observed for groups forming the hydrogen bonds.

SAQ 4.3c

Examine the infrared spectra of ethanol in Figures 4.3d and 4.3e. Figure 4.3d is of a 10 vol% solution of ethanol in CCl_4, run at a pathlength of 0.1 mm, while Figure 4.3e is a 1 vol% solution of the same compound in the same solvent at a pathlength of 1.0 mm. Can you account for any observed differences?

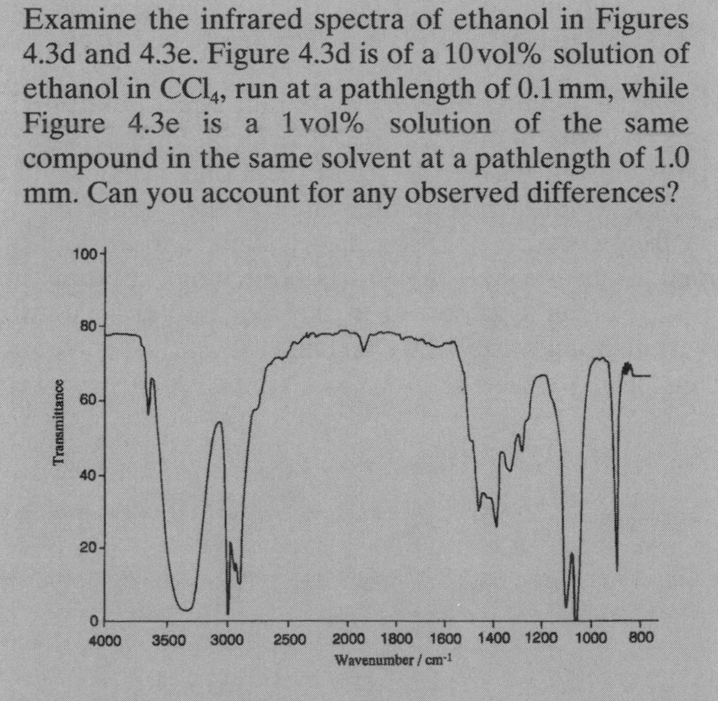

Fig. 4.3d. *The infrared spectrum of ethanol (10 vol % in CCl_4, 0.1 mm pathlength cell)*

SAQ 4.3c
(cont.)

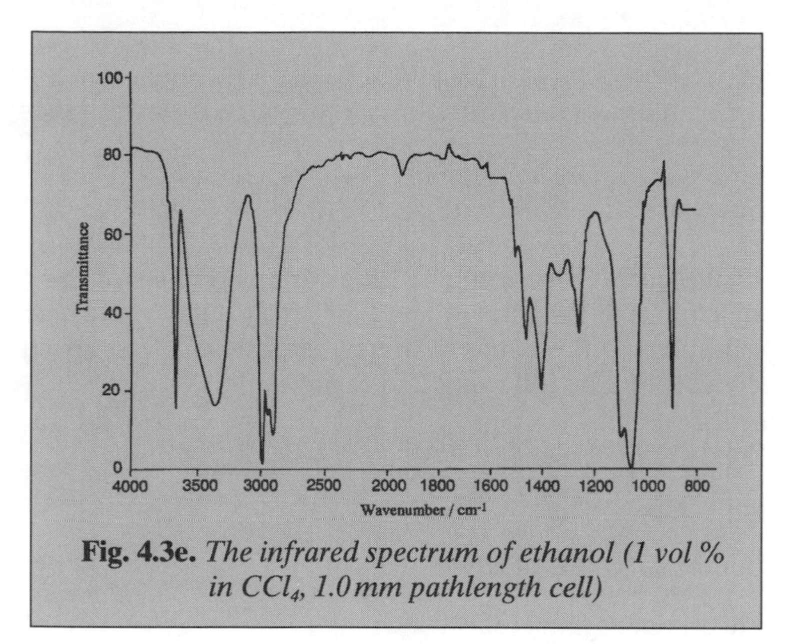

Fig. 4.3e. *The infrared spectrum of ethanol (1 vol % in CCl_4, 1.0 mm pathlength cell)*

Summary

In this chapter we examined the various regions of the mid-infrared spectrum, highlighting and assigning the particular molecular vibrations to the bands that may be observed. The spectrum may be divided into the following four regions:

X—H stretching region (4000–2500 cm^{-1});

triple bond region (2500–2000 cm^{-1});

double bond region (2000–1500 cm^{-1});

fingerprint region (1500–600 cm^{-1}).

A number of the factors which complicate the appearance, and hence the interpretation, of spectra were then presented. These included the following:

overtone and combination bands;

Fermi resonance;

hydrogen bonding.

Objectives

On completion of this chapter you should be able to:

- recognise the most reliable absorption frequencies for a particular functional group;
- use group frequencies to distinguish the spectra of samples;
- understand the factors which complicate infrared spectra and be able to recognise the effects that these have on the spectra.

5. Quantitative Analysis

5.1. INTRODUCTION

Quantitative infrared spectroscopy suffers certain disadvantages when compared with other analytical techniques and it tends to be confined to specialist applications. However, there are certain applications where it is used because it can be cheaper or faster. The technique is often used for the analysis of one component of a mixture, especially when the compounds in the mixture are alike chemically or have very similar physical properties (for example, structural isomers). In these instances analysis by using ultraviolet/visible spectroscopy is difficult because the spectra of the components will be nearly identical. Chromatographic analysis may be of limited use because the separation of isomers, for example, is difficult to achieve. The infrared spectra of isomers are usually quite different in the fingerprint region. Another advantage of the infrared technique is that it is non-destructive and requires only a small amount of sample.

In this chapter we will look at how infrared spectroscopy can be used for quantitative analysis. First, we will examine the various ways in which an infrared spectrum can be manipulated for analysis. We will also see how the Beer–Lambert Law can be applied to the analysis of both simple and multicomponent mixtures.

5.2. SPECTRUM MANIPULATION

There are a number of techniques available to users of modern infrared spectrometers which help with both the qualitative and quantitative interpretation of spectra.

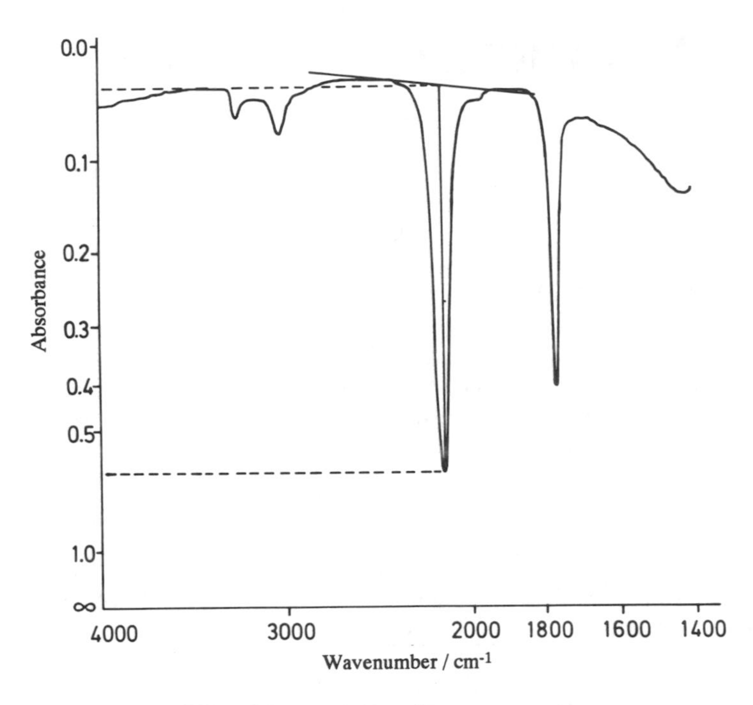

Fig. 5.2a. *A baseline correction*

5.2.1. Baseline Correction

It is usual in quantitative infrared spectroscopy to use a baseline joining the points of lowest absorbance on the peak, preferably in reproducibly flat parts of the absorption line. The absorbance difference between the baseline and the top of the band is then used. A baseline correction is shown in Figure 5.2a.

5.2.2. Smoothing

Noise in a spectrum can be diminished by smoothing. After a spectrum is smoothed it becomes similar to the result of an experiment at lower resolution. The features are blended into each other and the noise level decreases. A smoothing function is basically a convolution between the spectrum and a vector whose points are determined by the degree of smoothing you choose to apply. Generally, you will be asked to enter a degradation factor, which will be some positive integer. A low value, for example, one, will produce only subtle changes, while a high value has a more pronounced effect on your spectrum.

5.2.3. Derivatives

Spectra can also be differentiated. Figure 5.2b shows a single absorption peak and its first and second derivative.

Derivative techniques have long been used in quantitative ultraviolet/visible spectroscopy, with the benefits of the technique

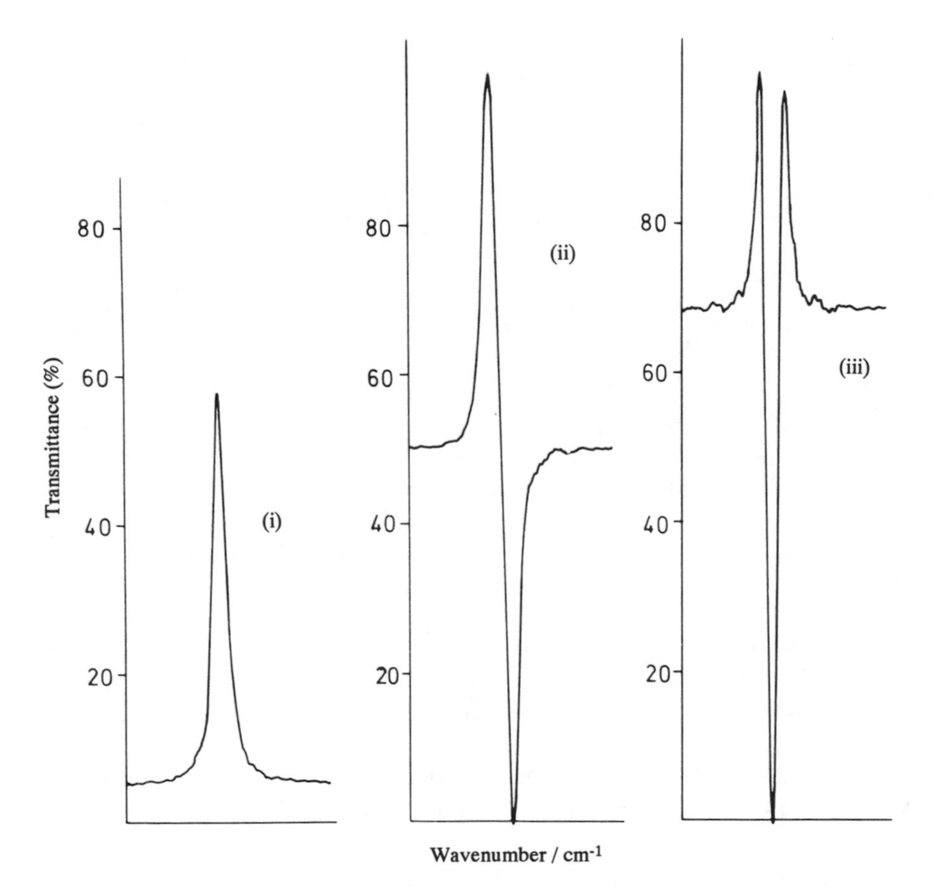

Fig. 5.2b. *Differentiation of spectra: (i) a single absorption peak; (ii) the first derivative; (iii) the second derivative*

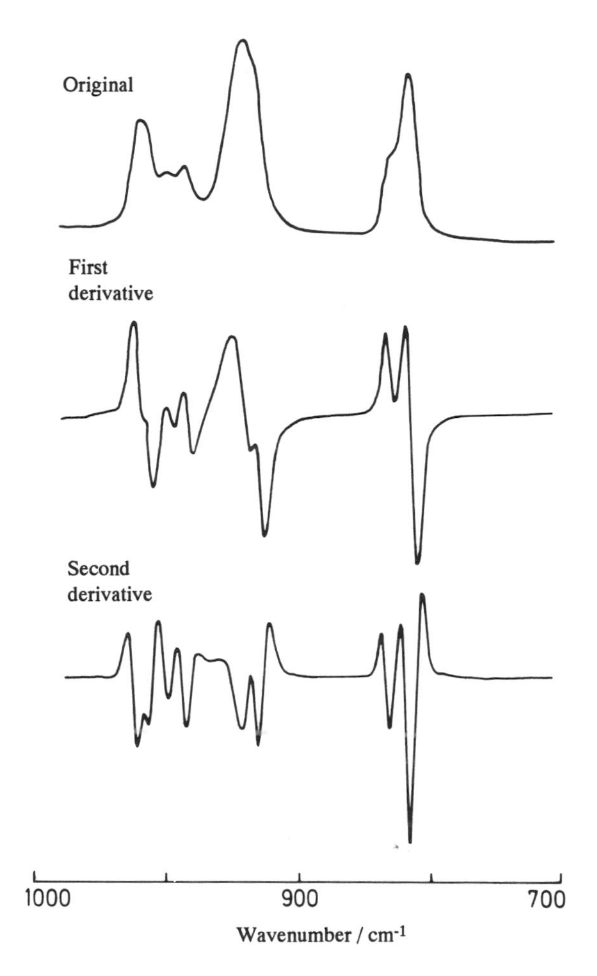

Fig. 5.2c. *A complex absorption band with first and second derivatives*

being twofold. Resolution is enhanced in the first derivative since we are looking at changes in gradient. The second derivative gives a negative peak for each band and shoulder in the absorption spectrum.

The advantage of derivatisation is more readily appreciated for more complex spectra and Figure 5.2c shows how differentiation can be used to resolve and locate peaks in an envelope. Note that sharp bands are enhanced at the expense of broad ones and this may allow selection of a peak even when there is a broad band beneath it. The latter point is clearly demonstrated in Figure 5.2d.

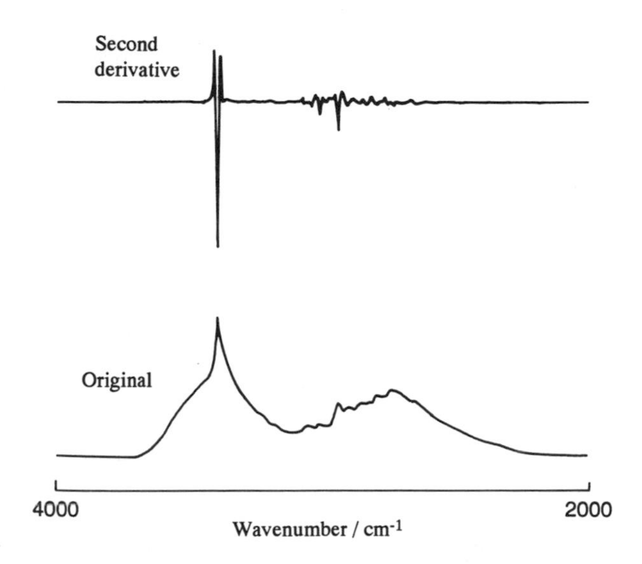

Fig. 5.2d. *The resolutition of a sharp peak from a broad background using derivative spectroscopy*

Often with modern FT-IR spectrometers it is possible to apply what is known as Fourier derivation. During this process the spectrum is transformed into an interferogram. It is then multiplied by an appropriate weighting function and finally it is retransformed to give the derivative. This technique provides more sensitivity.

5.2.4. Deconvolution

Deconvolution is the process of compensating for the intrinsic linewidths of modes in order to resolve overlapping bands. The technique yields spectra that have much narrower bands and is able to distinguish closely spaced features. The instrumental resolution is not increased, but the ability to differentiate spectral features can be significantly improved. This is illustrated by Figure 5.2e, which shows a broad band before and after deconvolution has been applied. You can see that peaks at quite close frequencies are now easily distinguished.

The deconvolution technique generally involves several steps:

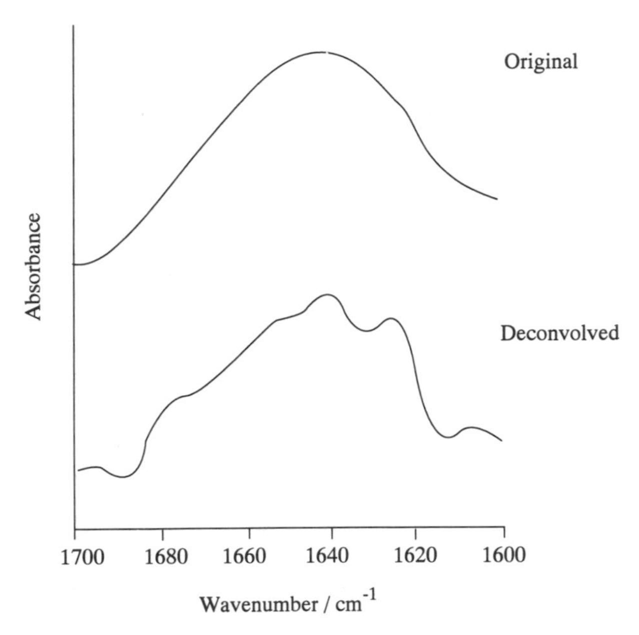

Fig. 5.2e. *A broad infrared band before and after deconvolution*

(i) computation of an interferogram of the sample by computing the inverse Fourier transform of the spectrum;

(ii) multiplication of the interferogram by a smoothing function and by a function consisting of a Gaussian–Lorentzian bandshape;

(iii) Fourier transformation of the resulting interferogram.

The deconvolution procedure is typically repeated iteratively for best results. At iteration, the lineshape is adjusted in an attempt to provide narrower bands without excessive distortion There are three parameters that can be adjusted to tune the lineshape:

(i) Proportion of Gaussian and Lorentzian lineshape. The scale of these components is adjusted depending on the predicted origins of the band shape. For example, it will vary depending on whether you are examining a solid, a liquid, or a gas.

(ii) Half-width. This is the width of the lineshape. The half-width is normally the same as or larger than the intrinsic linewidth (full width at half height) of the band. If the value specified is too small, the spectrum tends to show only small variations in intensity. If it is too large, distinctive negative side lobes are produced.

(iii) Narrowing function. This allows the bandwidths to be narrowed by varying degrees. Care must be exercised because if the function specified is too small, the resulting spectrum will not be significantly different from the original broad band. Likewise, if the narrowing function is too large, the spectrum may produce false peaks as noise starts to be deconvolved.

5.2.5. Curve-fitting

Quantitative values for band areas of heavily overlapped bands can be achieved by using curve-fitting procedures. Many curve-fitting procedures are based on a least-squares minimisation process. Least-squares curve-fitting covers a general class of techniques whereby one attempts to minimise the sum of the squares of the difference between an experimental spectrum and a computed spectrum generated by summing the component curves. Generally, the procedure involves entering the values of the frequencies of the component bands (determined using derivatives and/or deconvolution) and the program then determines the best estimate of the parameters of the component curves.

Apart from the obvious variables of peak height and width, the type of bandshape needs to be considered. The class of bandshape of an infrared spectrum depends on the type of sample. A choice of Gaussian, Lorentzian or a combination of these bandshapes is usually applied. Figure 5.2f illustrates the curve-fitting process.

5.3. DETERMINATION OF CONCENTRATION

Lambert, in the eighteenth century, found that the amount of light transmitted by a solid sample was dependent on the thickness of the sample. This was extended to solutions by Beer, during the following

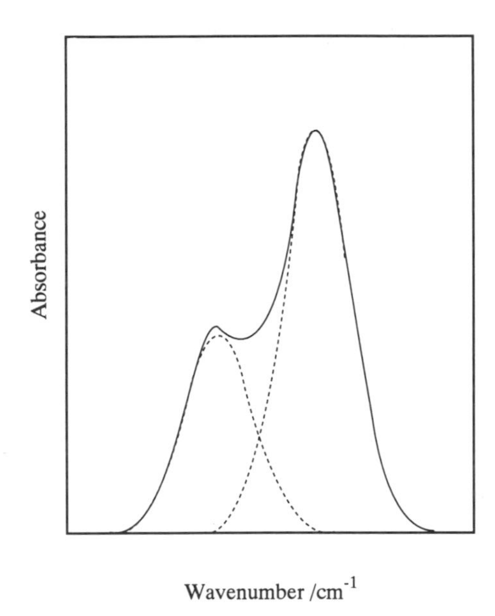

Fig. 5.2f. *Curve-fitting of overlapping infrared bands*

century. The resulting Beer–Lambert Law can be derived theoretically and applies to all electromagnetic radiation.

The *absorbance* of a solution is directly proportional to the thickness and the concentration of the sample as follows:

$$A = \epsilon c l$$

where A is the absorbance of the solution, c is the concentration and l is the pathlength of the sample. The constant of proportionality is usually given the symbol epsilon (ϵ), and is referred to as the *molar absorptivity*.

The absorbance can be shown to be equal to the difference between the logarithms of the intensity of the light entering the sample (I_0) and the intensity of the light transmitted (I) by the sample:

$$A = \log_{10} I_0 - \log_{10} I = \log_{10} (I_0/I)$$

Absorbance is therefore dimensionless.

Transmittance is defined by the following relationship:

$$T = I/I_0$$

and % transmittance as:

$$\% T = 100 \times T$$

It therefore follows that:

$$A = -\log_{10}(I/I_0) = -\log_{10} T$$

We can also write the following equation:

$$A = -\log_{10}(\% T/100) = \epsilon cl$$

When using % transmittance values it is easy to relate and to understand the numbers. For example, 50% transmittance means that half the light is transmitted and half is absorbed, while a value of 75% means that three quarters of the light is transmitted and one quarter absorbed.

Π What would be the absorbance of a solution which had a % transmittance of:

(a) 100%; (b) 50%; (c) 10%; (d) 0%?

Substituting these values into the above equation gives the following:

(a) $A = -\log_{10}(100/100) = 0$;

(b) $A = -\log_{10}(50/100) = 0.303$;

(c) $A = -\log_{10}(10/100) = 1.0$;

(d) $A = -\log_{10}(0/100) =$ infinity.

The important point to gain from these figures is that at an absorbance of 1.0, 90% of the light is being absorbed, so the instrument's detector does not have much radiation to work with.

SAQ 5.3

A 1.0% w/v solution of hexan-1-ol has an absorbance of 0.37 at 3660 cm^{-1} in a 1.0 mm cell. Calculate its molar absorptivity at this frequency.

The Beer–Lambert Law tells us that a plot of absorbance against concentration should be linear, with a gradient of ϵl, and passes through the origin. In theory, to analyse a solution of unknown concentration you need to prepare solutions of known concentration, choose a suitable peak, measure the absorbance at this frequency, and then plot a graph (a calibration graph). The concentration of the compound in solution can now be read, given its absorbance.

However, there are a few factors which need to be considered first when taking this approach:

(a) Preparation of solutions of known concentration.

The concentrations have to give us sensible absorbance values — not too weak and not too intense.

(b) Choosing a suitable absorption peak.
We would like this technique to be as sensitive as possible so we should choose an intense peak. However, infrared spectra often have many, sometimes overlapping, peaks. We need to find a peak isolated from the others, with a high molar absorptivity. A further problem that sometimes arises, especially in the spectra of solid samples, is the presence of asymmetric bands. In these cases, peak height cannot be used because the baseline will vary from sample to sample. Peak area measurements should be used instead. Most FT-IR spectrometers have accompanying software which can carry out these calculations.

(c) Measurement of absorbance.
Quantitative measurements need to be carried out on absorbance spectra. Thus, transmittance spectra need to be converted to absorbance spectra. Most spectrometers have this simple process incorporated into their software.

5.4. SIMPLE ANALYSIS

5.4.1. Analysis of Liquid Samples

We shall first look at the determination of a trace component in a two-component mixture. This can often be difficult using other analytical methods. The success of this method depends on the presence of an intense band due to the impurity.

Commercial propan-2-ol often contains traces of acetone formed by oxidation:

$$CH_3CH(OH)CH_3 \xrightarrow{[O]} CH_3COCH_3$$

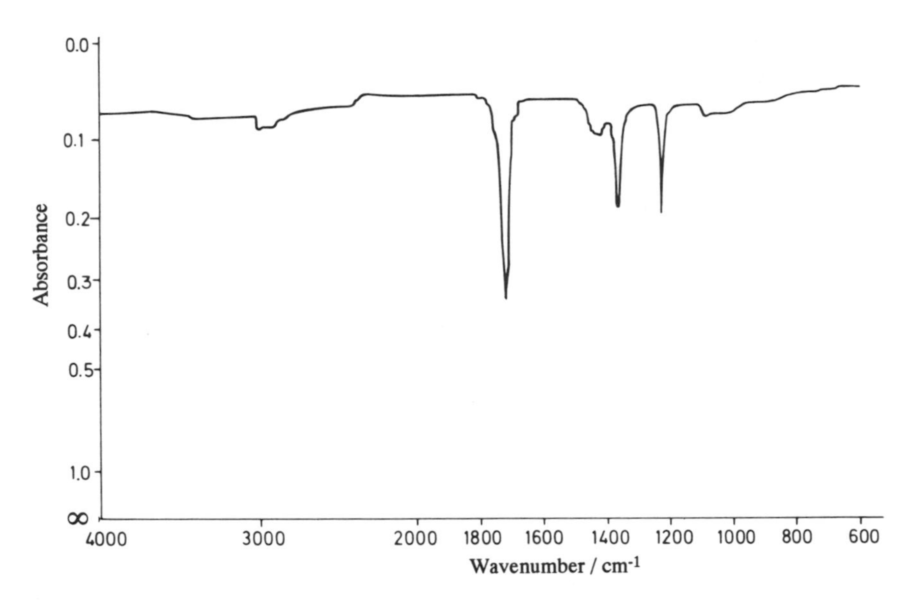

Fig. 5.4a. *An infrared spectrum of acetone*

Figure 5.4a is a spectrum of pure acetone, while Figure 5.4b is a spectrum of pure propan-2-ol.

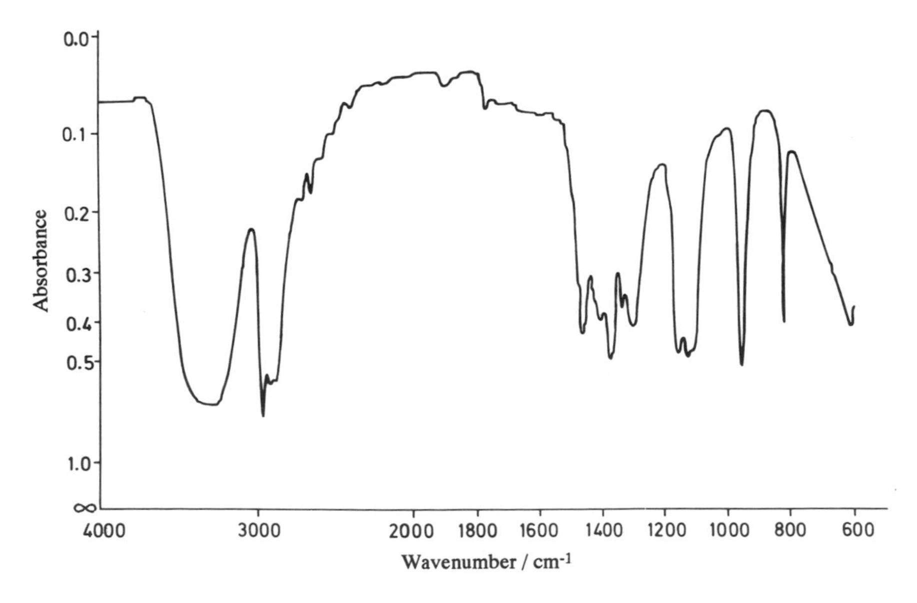

Fig. 5.4b. *An infrared spectrum of propan-2-ol*

There are two ways to approach this problem. We can either use a peak due to acetone or one from propan-2-ol. In this particular case, where we are interested in the low concentration of acetone, there will be little change in peak intensities from the propan-2-ol as its concentration varies from, e.g. 95 to 100%. Therefore, clearly we need a suitable peak from acetone and it is a matter of choosing the one which gives the best accuracy.

The chosen peak should:

- have a high molar absorptivity;
- not overlap with other peaks from other components in the mixture or solvent;
- be symmetrical;
- give a linear calibration plot of absorbance versus concentration.

The best peak to choose in this example is the C—O stretching absorption of acetone because it is an intense peak and lies in the region where no absorptions occur from the other component of the mixture.

The next step is to draw a calibration plot of absorbance against concentration.

SAQ 5.4a

Draw a plot of absorbance against concentration from the data given below and calculate the molar absorptivity in units of $m^2\ mol^{-1}$.

Acetone in CCl_4	Absorbance at 1719 cm^{-1}
(vol %)	
0.25	0.183
0.50	0.315
1.00	0.570
1.50	0.796
2.00	1.000

→

SAQ 5.4a
(cont.)

The absorbance values that are given were read straight from the spectrum. The baseline had an absorbance of 0.06 at 1719 cm^{-1}. The density of acetone is 0.790 g cm^{-3} and the pathlength was 0.1 mm.

SAQ 5.4a

The next stage in the analysis is to look at the spectrum of a sample of propan-2-ol and determine the amount of acetone present.

SAQ 5.4b

The infrared spectrum of a 10 vol% solution of commercial propan-2-ol in CCl_4 in a 0.1 mm pathlength cell is shown in Figure 5.4c.

→

SAQ 5.4b
(cont.)

(i) Determine the concentration (in mol dm^{-3}) of acetone in this solution by using the calibration curve plotted in SAQ 5.4a.

(ii) Calculate the % acetone in the propan-2-ol.

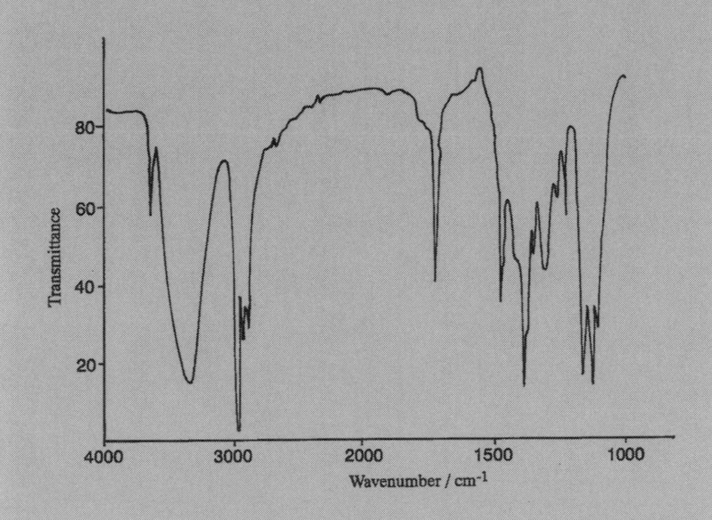

Fig. 5.4c. *An infrared spectrum of commercial propan-2-ol in carbon tetrachloride*

SAQ 5.4b

Infrared spectroscopy can be used to measure the number of functional groups in a molecule, for example the number of —OH or —NH_2 groups. It has been found that the molar absorptivity of the bands corresponding to the group is proportional to the number of groups, i.e. each group has its own intensity which does not vary drastically from molecule to molecule. This approach has been used to measure chain length in hydrocarbons by using the C—H deformation, the methylene group at 1467 and 1305 cm^{-1} and the number of methyl groups in polyethylene.

5.4.2. Analysis of Solid Samples

Solid mixtures can also be analysed. They are more susceptible to errors because of scattering. These analyses are usually carried out with KBr discs or in mulls. The problem here is the difficulty in measuring the pathlength. However, this measurement becomes unnecessary when an internal standard is used. Addition of a constant known amount of an internal standard is made to all samples and calibration standards.

The calibration curve is obtained by plotting the ratio of the absorbance of the analyte to that of the internal standard, against the concentration of the analyte. The absorbance of the internal standard varies linearly with the sample thickness and thus compensates for it. The discs or mulls must be made under exactly the same conditions to avoid intensity changes or shifts in band positions.

The standard must be carefully chosen and ideally, it should:

- have a simple spectrum with very few bands;
- be stable to heat and not absorb moisture;

- be easily reduced to a particle size less than the incident radiation without lattice deformation;

- be non-toxic, giving clear discs in a short time;

- be readily available in the pure state.

Some common standards that are used include calcium carbonate, sodium azide, napthalene and lead thiocyanate.

5.5. MULTICOMPONENT ANALYSIS

The analysis of a component in a complex mixture presents special problems. In this section we will examine how to determine the concentrations of a number of components in a solution by using xylene as an example. Commercial xylene is a mixture of isomers,

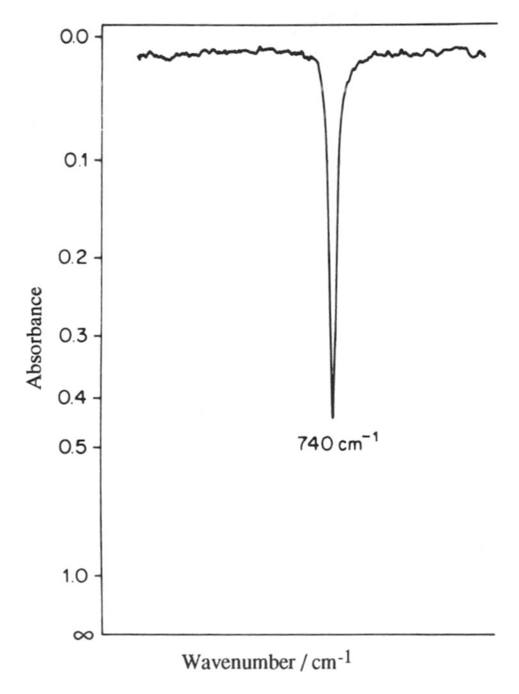

Fig. 5.5a. *The infrared spectrum of* o-*xylene in cyclohexane (1 vol%, 0.1 mm pathlength)*

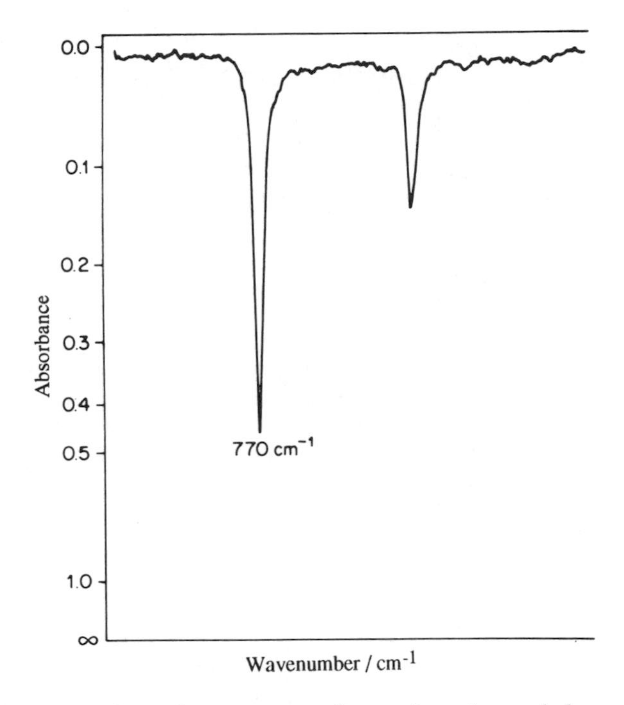

Fig. 5.5b. *The infrared spectrum of* m-*xylene in cyclohexane (2 vol%, 0.1 mm pathlength)*

namely 1,2-dimethylbenzene (*o*-xylene), 1,3-dimethylbenzene (*m*-xylene) and 1,4-demethylbenzene (*p*-xylene).

The spectra of these three pure xylenes in cyclohexane solution, Figures 5.5a–5.5c, all show strong bands in the 800–600 cm^{-1} region. Cyclohexane has a very low absorbance in this region and is therefore a suitable solvent for the analysis. The infrared spectrum of a commercial sample of xylene is given in Figure 5.5d.

We can estimate the concentration of the three isomers in the commercial sample. First, we need to measure the absorbencies of the xylenes at 740, 770 and 800 cm^{-1} by using the standards shown in Figures 5.5a–5.5c.

o-xylene 740 cm^{-1} $A = 0.440 - 0.012 = 0.428$

m-xylene 770 cm^{-1} $A = \dfrac{0.460 - 0.015}{2} = 0.223$

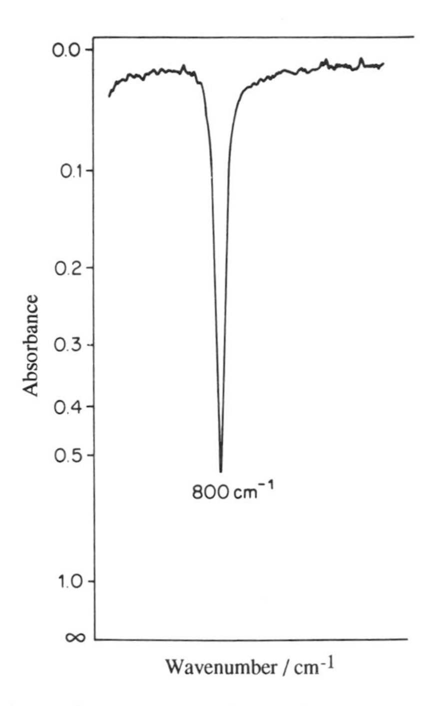

Fig. 5.5c. *The infrared spectrum of* p-*xylene in cyclohexane (2 vol%, 0.1 mm pathlength)*

p-xylene 800 cm^{-1} $A = \dfrac{0.545 - 0.015}{2} = 0.265$

The values for m-xylene and p-xylene are divided by two since these solutions are twice as concentrated. The above absorbance values are proportional to the molar absorptivity; hence the concentrations of the xylenes can be estimated in the mixture once the absorbencies are measured. From Figure 5.5d we find the following:

740 cm^{-1} $A = 0.194 - 0.038 = 0.156$

770 cm^{-1} $A = 0.720 - 0.034 = 0.686$

800 cm^{-1} $A = 0.133 - 0.030 = 0.103$

Dividing these absorbance values by the standard values above gives the vol% of each isomer in the mixture:

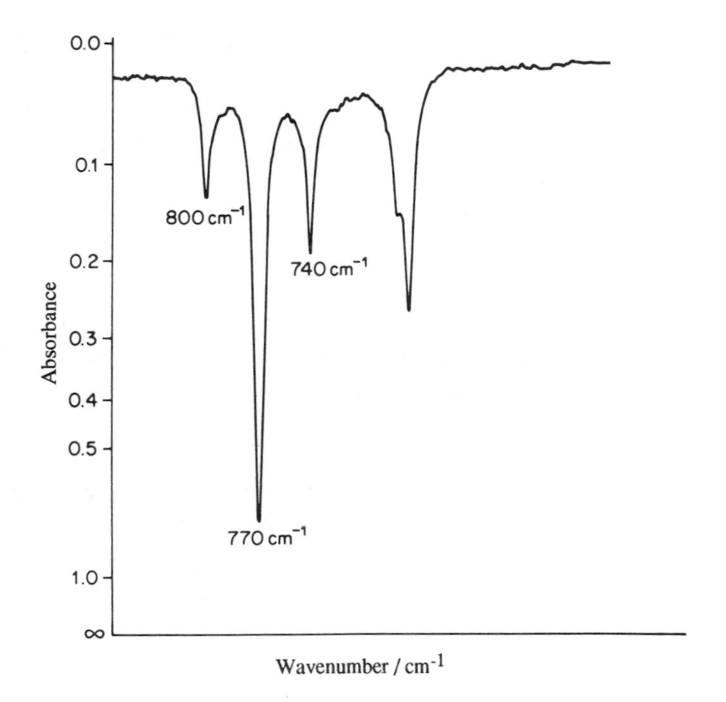

Fig. 5.5d. *The infrared spectrum of commercial xylene in cyclohexane (5 vol%, 0.1 mm pathlength)*

o-xylene = 0.156/0.428 = 0.364 vol%

m-xylene = 0.686/0.223 = 3.076 vol%

p-xylene = 0.103/0.265 = 0.389 vol%

It should be pointed out that there could be a certain amount of error in these values. There could be error due to the fact that it is difficult to select a baseline for the analysis because of overlapping peaks. We have also ignored any non-linearity of the Beer Law plots, having used only one value to determine the constant of proportionality above for each solution.

Alternatively, this same problem of xylene can be determined by spectral subtraction. If a sample of the impurity is available, mixtures of the starting material and product can be separated by subtraction of the spectra of the mixture and impurities. This is achieved by subtracting a fraction of *o*-xylene, then *m*-xylene, and finally *p*-xylene, from the commercial xylene spectrum to null the corresponding absorptions. The subtraction factors can be read from the spectrometer computer. This second method is of course much faster.

Summary

In this chapter we began by looking at the various ways in which a spectrum can be manipulated in order to carry out quantitative analysis. These included baseline correction, smoothing, derivatives, deconvolution and curve-fitting.

The Beer–Lambert Law was also introduced, showing how the intensity of an absorption band is related to the amount of analyte present. This was then applied to the simple analysis of liquid and solid samples, followed by treatment of multicomponent mixtures.

Objectives

On completion of this chapter you should be able to:

- describe and apply a variety of analytical techniques used for compensating for background absorption and overlapping peaks;
- convert transmittance values to the corresponding absorbance values;
- use the Beer–Lambert Law in quantitative analysis;
- plot calibration graphs of absorbance against concentration and analyse simple mixtures;
- analyse multicomponent systems.

6. Applications

6.1. INTRODUCTION

There are numerous types of materials which can be examined using infrared spectroscopy. Organic and inorganic molecules, polymers, biological molecules, pollutants, drugs, fibres, paints, oils and lubricants, agricultural chemicals, food additives, catalysts, flame retardants, forensic materials, minerals, clays, organometallics, petroleum products, etc., have all been studied by using infrared spectroscopy. For most of the above materials, extensive special libraries exist for reference purposes.

You will now have an understanding of how an infrared spectrometer works and how to prepare a range of sample types. You should also be able to interpret a spectrum and, if necessary, analyse a spectrum quantitatively. In this chapter you can now apply this knowledge to a variety of systems.

6.2. ORGANIC COMPOUNDS

One of the most common applications of infrared spectroscopy is for the identification of organic compounds. The major classes of organic molecules are examined in turn in this section and useful group frequencies are detailed. The information here should provide a good reference source of information at a future date.

6.2.1. Alkanes

These compounds contain only C—H and C—C bonds. The C—C stretching vibration occurs in the fingerprint region, is usually very weak and couples with other C—C stretching vibrations. It is therefore of little use diagnostically. In contrast C—H stretching is a useful absorption, occurring in saturated hydrocarbons below 3000 cm^{-1}, and is usually of medium intensity. The C—H stretching absorptions of alkanes are summarised here:

Frequency (cm^{-1})	Assignment
2962 (medium)	Asymmetrical C—H stretching, CH_3
2872 (medium)	Symmetrical C—H stretching, CH_3
2926 (medium)	Asymmetrical C—H stretching, CH_2
2853 (medium)	Symmetrical C—H stretching, CH_2
2890 (weak)	C—H stretching, CH
1380	Symmetrical C—H deformation, CH_3
1465	C—H deformation, CH_2
1450	Asymmetrical C—H deformation, CH_3
720	Rocking and wagging, $(CH_2)_n$ with $n>4$

SAQ 6.2a

Examine the spectrum of nonane in Figure 6.2a and describe the vibrations corresponding to the absorptions marked A, B and C.

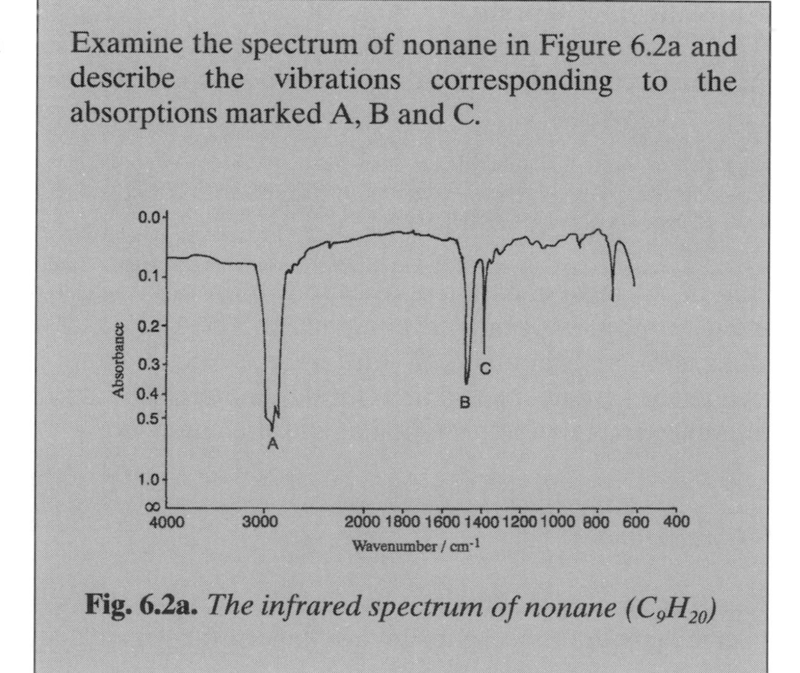

Fig. 6.2a. *The infrared spectrum of nonane* (C_9H_{20})

SAQ 6.2a

6.2.2. Alkenes

These compounds contain the C=C group, with the majority having hydrogen attached to the double bond. Four vibrational modes are associated with this molecular fragment. These are the out-of-plane and in-plane C—H deformations, C=C stretching and =C–H stretching. The modes observed for alkenes are listed below:

Frequency (cm^{-1})	Assignment
3100–3000	=C–H stretching
1680–1600	C=C stretching
1400	C—H in-plane deformation
1000–600	C—H out-of-plane deformation

6.2.3. Alkynes

Alkynes contain the C≡C group and three characteristic absorptions can be present, namely C—H stretching, C—H bending and C≡C stretching, depending on the structure of the compound. These absorptions are as follows:

Frequency (cm^{-1})	Assignment
3320–3220 (strong)	C—H stretching
2300–2100 (weak)	C≡C stretching
700–600	C—H bending

SAQ 6.2b

Figure 6.2b shows the spectrum of phenylethyne. Identify the C—H stretching and C≡C stretching modes in this spectrum.

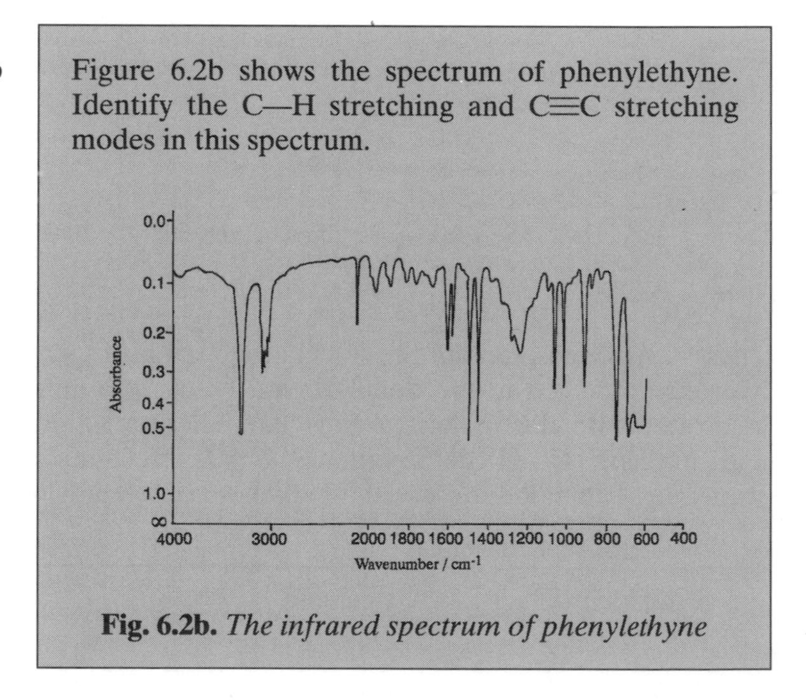

Fig. 6.2b. *The infrared spectrum of phenylethyne*

SAQ 6.2b

6.2.4. Aromatic Compounds

The presence of a benzene ring in a molecule gives rise to absorption in six regions of the spectrum, as follows:

Frequency (cm^{-1})	Assignment
3100–3000	C—H stretching
2000–1650 (weak)	Overtone and combination bands
1600–1550	Ring stretching
1500–1450	Ring stretching
1300–1000 (weak)	C—H in-plane bending
900–600 (strong)	C—H out-of-plane bending

The out-of-plane bending vibrations of aromatic compound bands are strong and characteristic of the number of hydrogens in the ring, and hence can be used to give the substitution pattern. This information is summarised in Figure 6.2c.

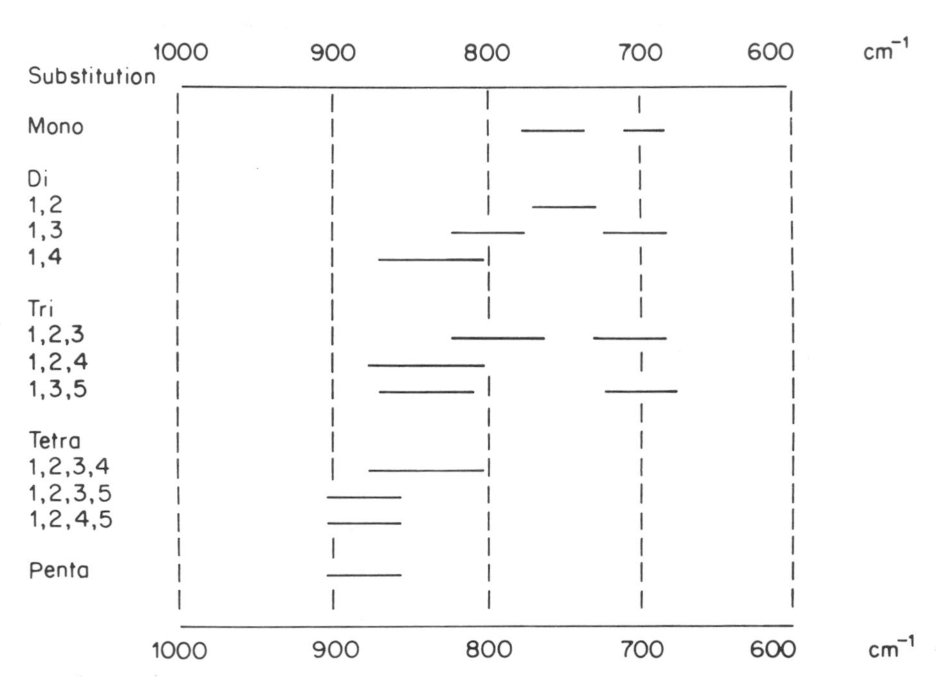

Fig. 6.2c. *The out-of-plane bending vibrations of substituted benzenes*

SAQ 6.2c

Examine the three spectra in Figures 6.2d–6.2f, which have been obtained for various isomeric disubstituted benzenes. Which is 1,2- which 1,3- and which is 1,4-disubstituted?

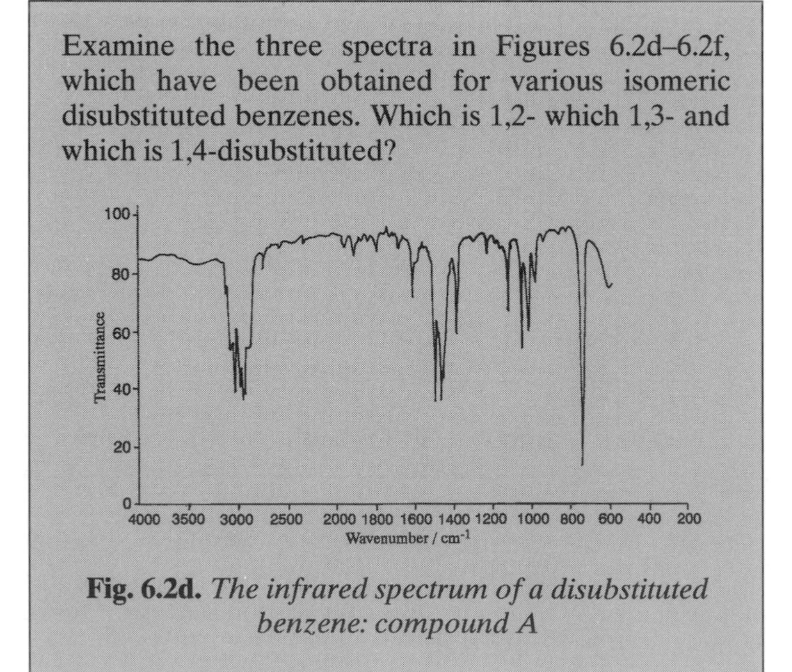

Fig. 6.2d. *The infrared spectrum of a disubstituted benzene: compound A*

SAQ 6.2c
(cont.)

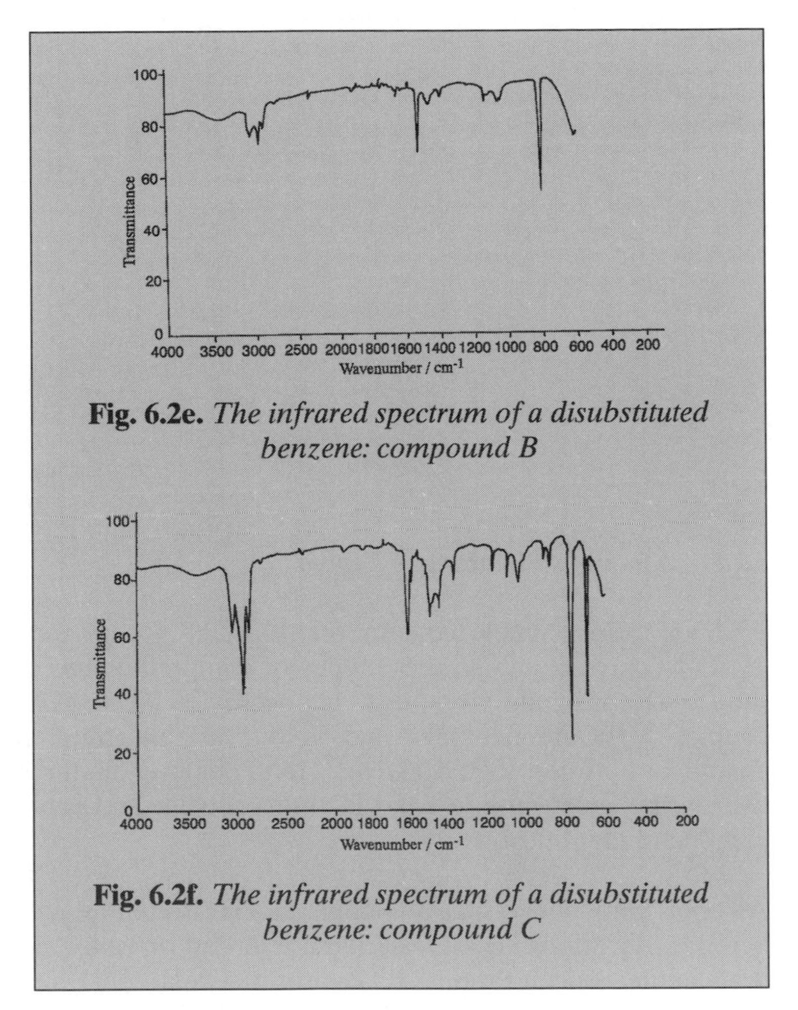

Fig. 6.2e. *The infrared spectrum of a disubstituted benzene: compound B*

Fig. 6.2f. *The infrared spectrum of a disubstituted benzene: compound C*

SAQ 6.2c

6.2.5. Alcohols, Phenols and Ethers

Alcohols and phenols contain the C—OH group and ethers the C—O—C group, and so these types of compounds therefore display the C—O stretching vibration. This occurs in the fingerprint region, couples with other modes and is of variable frequency, usually absorbing between 1300–1000 cm^{-1}. Its redeeming features is that it is usually the most intense band in the spectrum and can be used for compound identification.

Alcohols and phenols also contain O—H stretching and C—O—H bending vibrations. The O—H band is broad or weak, depending on the sample concentrations. It absorbs between 2500 and 3650 cm^{-1}, depending on molecular association, and gives valuable structural information. The C—O—H bending vibration occurs in the fingerprint region and is of little diagnostic use.

6.2.6. Amines

Amines are classified as primary (containing the NH_2 group), secondary (containing the NH group) and tertiary (containing no hydrogen attached to nitrogen). In addition to hydrocarbon frequencies, amines would be expected to give rise to N—H stretching and N—H bending absorptions as follows:

Frequency (cm^{-1})	Assignment
3490–3180	N—H stretching: 100 cm^{-1} doublet for primary amines; singlet for secondary amines; absent for tertiary amines
1650–1580	N—H bending, medium

6.2.7. Aldehydes and Ketones

These compounds contain the carbonyl group (C═O), with the following assignments:

Frequency (cm^{-1})	Assignment
1780–1660	Ketone C═O stretching
1740–1670	Aldehyde C═O stretching
2900–2700	Aldehyde C—H stretching

The position of the C═O stretching frequency within these ranges is dependent on hydrogen bonding and conjugation.

6.2.8. Other Carbonyl Containing Compounds

The major bands which appear in the infrared spectra of carboxylic acids (which contain the COOH group) are summarised below:

Frequency (cm^{-1})	Assignment
1715–1680	C=O stretching
3500–2500	O—H stretching
1300–1200	C—O stretching
1400	C—O—H in-plane bending
900	C—O—H out-of-plane bending

The two most polar bonds in esters (containing —CO—O—C—) are the C=O and C—O bonds and these bonds listed below produce the strongest peaks in the spectrum of any ester:

Frequency (cm^{-1})	Assignment
1400–1000	C—O stretching, strong
1750–1710	C=O stretching

Acid anhydrides (containing —CO—O—C—CO—) are usually easily distinguished from other carbonyl containing compounds because the C=O frequency is invariably a doublet, with assignments as follows:

Frequency (cm^{-1})	Assignment
1850–1730	C=O stretching, doublet
Fingerprint region	C—O stretching

6.2.9. Other Nitrogen Containing Compounds

The important infrared modes for amides (—CO—NH—), nitro compounds (—C—NO_2) and nitriles (—C≡N) are shown below:

Frequency (cm^{-1})	Assignment
	Amides
1700–1640	C=O stretching
3500–3100	N—H stretching
	Nitro compounds
1570–1450	N=O stretching
1370–1300	N=O stretching
	Nitriles
2250, medium	C≡N stretching

6.2.10. Sulphur Containing Compounds

The S—H stretching frequency is very weak and may even be too weak to observe. The oxygen containing sulphur compounds are more complex. The assignment of modes to a variety of sulphur containing compounds are listed overleaf:

Frequency (cm^{-1})	Assignment
	Sulphoxides
1100–1000	S=O stretching, strong
	Sulphones, sulphonic acids and sulphonamides
1160–1120	S=O stretching
1350–1300	S=O stretching
	Sulphonyl chlorides
1210–1150	S=O stretching
1400–1330	S=O stretching

6.2.11. Halogen Containing Compounds

The carbon–halogen stretching absorption is usually intense, but occurs at low frequencies, often outside the normal range, and can easily be confused with C—H bending frequencies from aromatic rings. The more massive the halogen, then the lower the frequency, as shown below:

Frequency (cm^{-1})	Assignment
1400–100	C—F stretching
800–600	C—Cl stretching

The C–I and C–Br stretching frequencies are normally outside the range of commonly used instruments.

6.3. INORGANIC COMPOUNDS

Inorganic compounds are commonly studied by using infrared spectroscopy. A useful reference on this subject is *Infrared and Raman Spectra of Inorganic and Coordination Compounds* by K. Nakamoto (Wiley, 1986) which provides extensive and detailed assignments of a range of inorganic materials. We will briefly look here at the assignments of some of the major classes of inorganic compounds.

6.3.1. Boron Compounds

Compounds containing the B—O linkage are characterised by a strong B—O stretching mode at 1380–1310 cm^{-1}. Boronic acid and boric acid contain OH groups and so show an O—H stretching mode in the region 3300–3200 cm^{-1}. Compounds containing a B—N group such as borazines and aminoboranes show a strong absorption in the region 1465–1330 cm^{-1}, due to B—N stretching.

The B–H stretching due to BH and BH_2 groups produce bands in the range 2640–2350 cm^{-1}. The BH_2 absorption is usually a doublet due to symmetric and asymmetric vibrations. There is also a deformation band in the region 1205–1140 cm^{-1} and a wagging vibration at 975–920 cm^{-1} due to the BH_2 group. In addition, a series of bands in the region 2220–1540 cm^{-1} is due to B···H···B bridge bonds.

Table 6.3a provides a summary of the main vibrational modes in boron compounds.

6.3.2. Silicon Compounds

The Si—H stretching mode produces an absorption in the region 2250–2100 cm^{-1}, while Si—H bending appears in the region 950–800 cm^{-1}. The Si—CH_3 group is characterised by a strong band at 1280–1255 cm^{-1} due to symmetric CH_3 deformation and an absorption due to methyl rocking and Si—C stretching at 860–760 cm^{-1}. The asymmetric CH_3 deformation absorbs only weakly near 1410 cm^{-1}. A medium intensity band due to the Si—CH_2—R group occurs at

Table 6.3a. *Characteristic infrared bands of boron compounds*

Frequency (cm^{-1})	Assignment
3300–3200	B—O–H stretching
2640–2350	B—H stretching
2220–1540	B···H···B bridging
1465–1330	B–N stretching
1380–1310	B–O stretching
1205–1140	B–H deformation
975–920	B–H wagging

1250–1200 cm^{-1}. An intense band appears at 1150 cm^{-1} for compounds containing the Si—C_6H_5 group.

The Si—O—R group produces at least one strong band at 1110–1000 cm^{-1} due to an asymmetric Si—O—C stretching, while Si—O stretching for the Si—O—C_6H_5 group absorbs in the range 970–920 cm^{-1}. Siloxanes produce a strong band at 1130–1000 cm^{-1} due to asymmetric Si—O—Si stretching. The Si—OH group absorbs in the range 3700–3200 cm^{-1}, and Si—O stretching for Si—OH produces a strong band at 910–830 cm^{-1}.

Some of the common infrared modes of silicon compounds are summarised in Table 6.3b.

6.3.3. Phosphorus Compounds

The P—H stretching mode gives rise to a medium intensity band in the range 2440–2275 cm^{-1}. Deformation due to the PH_2 group produces a medium intensity band at 1090–1080 cm^{-1} and the PH_2 wagging vibration gives rise to a band near 840–810 cm^{-1}.

The P═O bond produces a strong stretching mode in the region 1320–1140 cm^{-1}. Most organic phosphorus acids with a P—OH group give rise to a strong band at 1040–910 cm^{-1}. A strong band appears at 1000–870 cm^{-1} in the spectra of compounds containing the P—O—P group, due to P—O—P stretching.

Table 6.3b. *Characteristic infrared bands of silicon compounds*

Frequency (cm^{-1})	Assignment
3700–3200	Si—OH stretching
2250–2100	Si—H stretching
1280–1255	Si—CH_3 symmetric deformation
1250–1200	Si—CH_2—R stretching
1150	Si—C_6H_5
1130–1000	Si—O—Si asymmetric stretching
1110–1000	Si—O—R asymmetric stretching
970–920	Si—O—C_6H_5
950–800	Si—H bending
860–760	Si—C stretching

Table 6.3c summarises some of the infrared modes produced by phosphorus compounds.

Table 6.3c. *Characteristic infrared bands of phosphorus compounds*

Frequency (cm^{-1})	Assignment
2700–2550,2300–2100,1040–910	P—OH stretching
2960,1460,1190–1170	P—O—CH_3 stretching
2440–2275	P—H stretching
1450–1425, 1130–1090,1010–900	P—C_6H_5
1300–1140	P=O stretching
1310–1280,960–860	P—CH_3 stretching
1240–1160,995–860	P—O—C_6H_5
1110–930	P—N
1090–1080	PH_2 deformation
1050–970	P—O—C stretching (aliphatic)
1000–870	P—O—P stretching
890–720	P—F stretching
840–810	PH_2 wagging
750–580	P=S stretching
580–440	P—Cl stretching

SAQ 6.3

Figure 6.3 shows the spectrum of the following phosphorus compound:

CH_3 S

P

Cl Cl

Locate the P—CH_3, PCl_2 and P═S stretching vibrations in the spectrum shown in Figure 6.3.

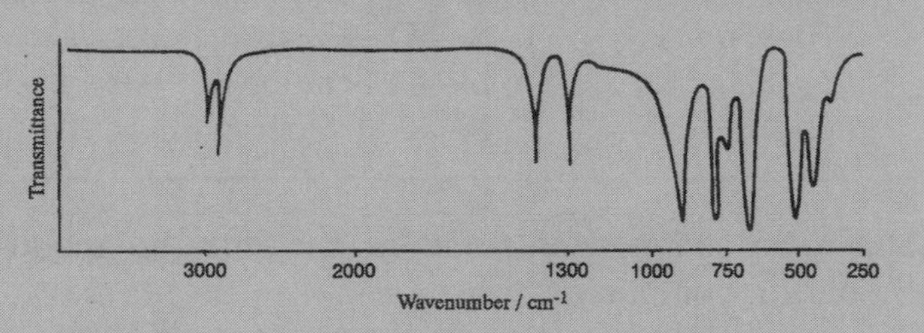

Fig. 6.3. *The infrared spectrum of an unknown phosphorus compound*

6.4. POLYMERS

Infrared spectroscopy is a popular method for characterising polymers, with most studies involving the determination of the molecular composition of the materials. In addition to providing valuable information about the chemical structures of a polymer, infrared spectroscopy can also provide information about the physical structure, which strongly influences the physical properties of the polymer. This aspect of infrared spectroscopy has been well developed for polymer systems, with many studies of configurational and conformational isomerism, hydrogen bonding, chain order and crystallinity of polymers.

6.4.1. Sample Preparation

There is a variety of methods available for examining polymer samples. If the polymer is a thermoplastic (for example, polyethylene or polystyrene) it can be softened by warming and pressed into a thin film in a hydraulic press. Alternatively, the polymer can be dissolved in a volatile solvent and the solution allowed to evaporate to a thin film on an alkali halide plate. Some polymers, such as crosslinked synthetic rubbers, can be microtomed (cut into thin slices with a blade). If the polymer is in the form of a surface coating, reflectance techniques can be used. In addition, a solution in a suitable solvent is a possibility.

6.4.2. Polypropylene

Polypropylene is a common polymer, used widely for packaging and moulded materials. Polypropylene provides a good example of the importance of *tacticity* in polymers. During the polymerisation process of this polymer three different *stereoisomers* are produced and these are shown in Figure 6.4a. In *isotactic* polymers all the repeat units have the same configuration, whereas in *syndiotactic* polymers the configuration alternates from one repeat unit to the next. *Atactic* polymers have random placement of the other configurations.

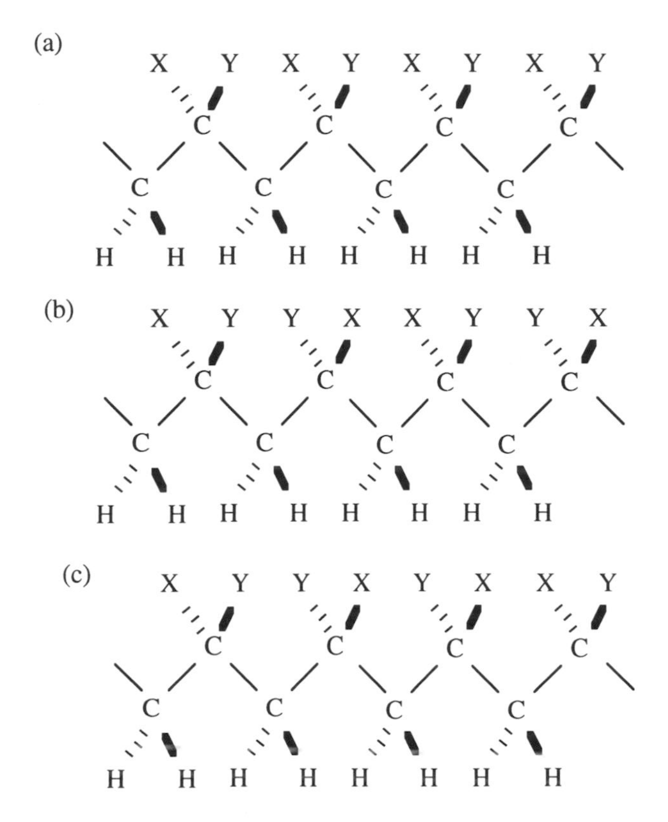

Fig. 6.4a. *The stereoisomers of polypropylene where X=H and Y=CH_3: (a) isotactic; (b) syndiotactic; (c) atactic*

Commercial polypropylene is essentially isotactic, and due to its regular structure, is crystalline and so gives rise to good mechanical properties. In contrast, atactic polypropylene is unable to crystallise because of its irregular structure and is a soft amorphous material which has no useful mechanical properties. The infrared spectra of the isotactic, syndiotactic and atactic forms of polypropylene display characteristic differences, as shown in Figure 6.4b. The absorptions at 970 and 1460 cm^{-1} do not depend upon the tacticity, whereas the absorptions at 840, 1000 and 1170 cm^{-1} are characteristic of isotactic polypropylene, and the absorption at 870 cm^{-1} is characteristic of syndiotactic polypropylene. Such differences are due to the different helical structures present in the isomers and can be used to estimate the fractions of isotactic and syndiotactic sequences in samples.

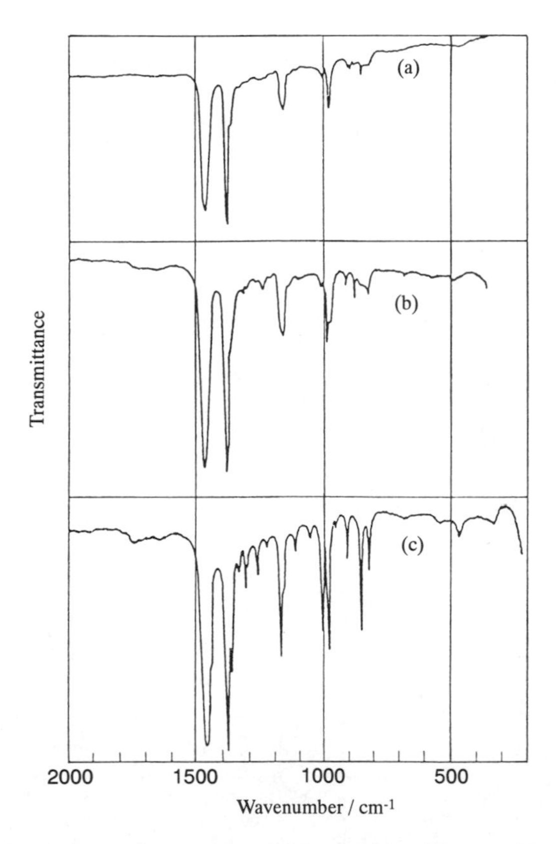

Fig. 6.4b. *The infrared spectra of (a) atactic, (b) syndiotactic and (c) isotactic polypropylene*

SAQ 6.4

The polymer poly(methyl methacrylate) (PMMA) has the following structural repeat unit:

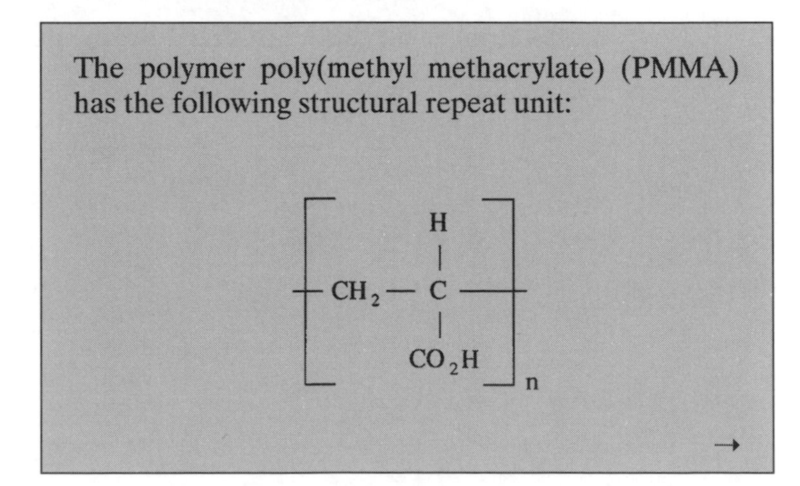

SAQ 6.4
(cont.)

The infrared spectrum of PMMA is shown in Figure 6.4c. Can you identify the C—C—O, C—H and C=O stretching, and OCH_3 bending modes in this spectrum?

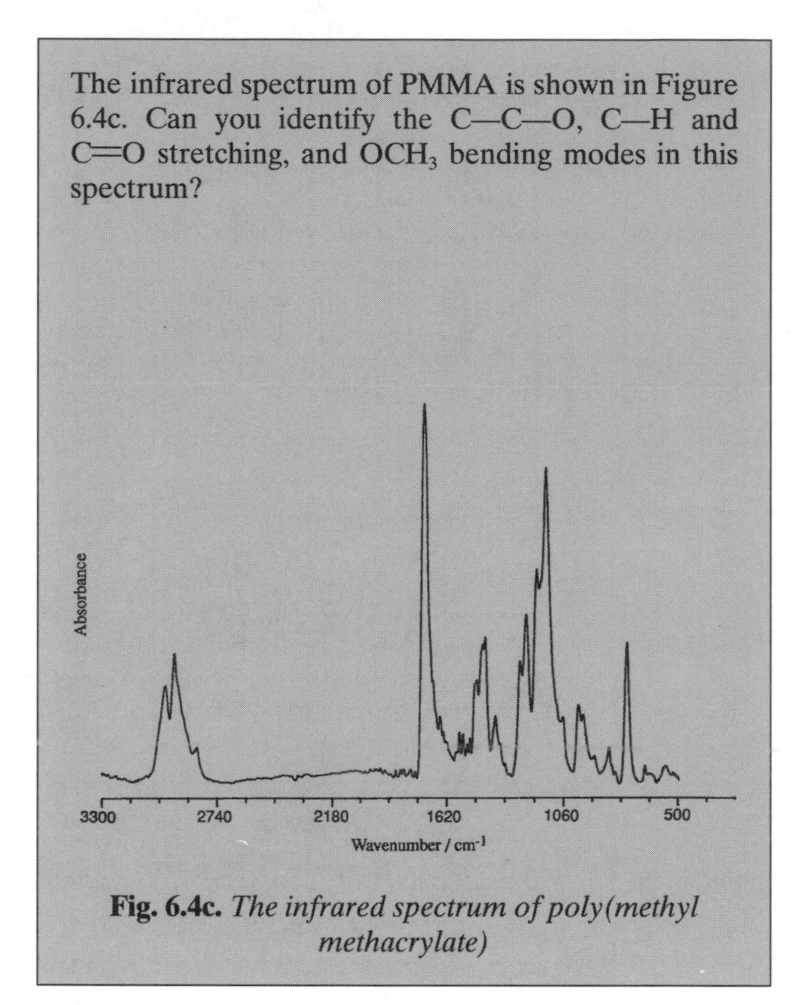

Fig. 6.4c. *The infrared spectrum of poly(methyl methacrylate)*

6.5. BIOLOGICAL SYSTEMS

An understanding of the relationship between the structure and function of biological materials is one of the most challenging areas in biology today. Infrared spectroscopy has proved to be a powerful tool for studying biological molecules and its application to biological problems is expanding. We will concentrate on two areas which have been successfully studied using infrared spectroscopy, namely proteins and lipids.

6.5.1. Sample Preparation

Water is a commonly used solvent for biological samples. However, NaCl cannot be used as a infrared window material in this case as it is very soluble in water, and CaF_2 or BaF_2 are better alternatives. Small pathlengths (~0.010 mm) are available and and help reduce the intensity of the very strong infrared modes produced in the water spectrum. The small pathlength also produces a small sample cavity, allowing samples in milligram quantities to be examined.

Certain difficulties arise when using water as a solvent in infrared spectroscopy. The infrared modes of water are very intense and may overlap with the sample modes of interest. This problem may be overcome by substituting water with deuterium oxide (D_2O). The infrared modes of D_2O occur at different frequencies to those observed for water because of the mass dependence of the vibrational frequency.

6.5.2. Proteins

The infrared spectra of proteins exhibit absorption bands associated with their characteristic amide group (CONH), which is the structural unit common to all molecules of this type. Characteristic bands of the amide group of protein chains are similar to absorption bands exhibited by secondary amides and are labelled as amide bands. There are nine such bands, called amide A, amide B and amide I–VII,

Table 6.5a *Characteristic infrared amide bands of proteins*

Designation	Approximate frequency (cm^{-1})	Nature of vibration
A	3300	N—H stretching in resonance
B	3110	with overtone (2 × amide II)
I	1653	80% C=O stretching; 10% C—N stretching; 10% N—H bending
II	1567	60% N—H bending; 40% C—N stretching
III	1299	30% C—N stretching; 30% N—H bending; 10% C=O stretching; 10% O=C—N bending; 20% other
IV	627	40% O=C—N bending; 60% other
V	725	N—H bending
VI	600	C=O bending
VII	200	C—N torsion

in order of decreasing frequency, and these are summarised in Table 6.5a.

The most useful infrared band for the analysis of the secondary structure of proteins in aqueous solution is the amide I band, occurring between approximately 1600 and 1700 cm^{-1}. The exact frequency of this vibration depends on the nature of the hydrogen bonding involving the C=O and N—H groups, and this is determined by the particular secondary structure adopted by the protein. Proteins generally contain a variety of domains in different conformations. As a consequence, the observed amide I band is usually a complex composite, consisting of a number of overlapping component bands representing the different types of protein secondary structure, namely α-helices, β-sheets, turns and non-ordered structures.
The method used for the estimation of protein secondary structure involves curve-fitting the amide I band. We looked at this quantitative method in Chapter 5. The parameters required, plus the number of component bands and their positions, are obtained from deconvolved

and derivative spectra (also discussed in Chapter 5). The fractional areas of the fitted component bands are directly proportional to the relative proportions of the structures that they represent. The percentages of α-helix, β-structure and turns are then estimated by addition of the areas of all of the component bands assigned to each of these structures and expressing the sum as a fraction of the total amide I area. An example of this approach is illustrated by Figure 6.5, which shows the amide I mode of the enzyme lysozyme.

6.5.3. Lipids

Lipids are insoluble organic compounds found in biological tissues, e.g. fats, which form the basic framework of all biological membranes. The infrared spectra of membrane lipids, for instance, can be divided into spectral regions which originate from the molecular vibrations of the hydrocarbon tail, the interface region and the head group. Representative frequencies of some important lipid bands are given in Table 6.5.

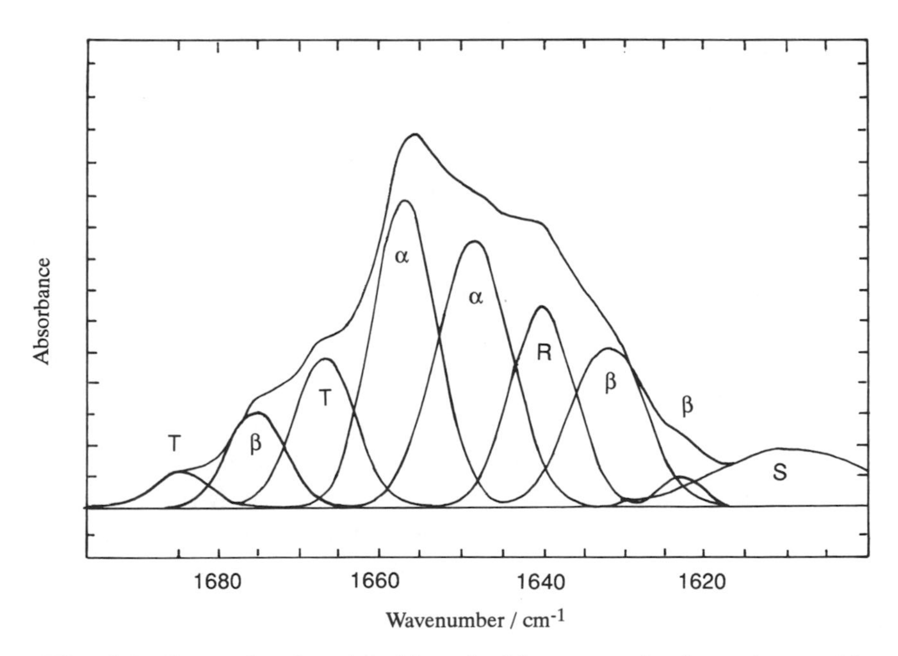

Fig. 6.5. *Curve-fitted amide I band of lysozyme in deuterium oxide*

Table 6.5b *Characteristic infrared bands of lipids*

Approximate frequency (cm^{-1})	Nature of vibration
3010	=C—H stretching
2956	CH_3 asymmetric stretching
2920	CH_2 asymmetric stretching
2870	CH_3 symmetric stretching
2850	CH_2 symmetric stretching
1730	C=O stretching
1485	$(CH_3)_3N^+$ asymmetric bending
1473, 1472, 1468, 1463	CH_2 scissoring
1460	CH_3 asymmetric bending
1405	$(CH_3)_3N^+$ symmetric bending
1378	CH_3 symmetric bending
1400–1200	CH_2 wagging band progression
1228	PO_2^- asymmetric stretching
1170	CO—O—C asymmetric stretching
1085	PO_2^- symmetric stretching
1070	CO—O—C symmetric stretching
1047	C—O—P stretching
972	$(CH_3)_3N^+$ asymmetric stretching
820	P—O asymmetric stretching
730,720,718	CH_2 rocking

Bands arising from the acyl chains (including CH_3 and CH_2 asymmetric and symmetric stretching vibrations, and CH_2 bending and rocking vibrations), the headgroup (PO_2^- stretching), and the interface region (C=O stretching) have all been used as conformational probes. Bands arising from the interfacial and headgroup region provide information concerning bilayer hydration. The ester C=O band is particularly useful, e.g. the deconvolved ester C=O stretching band of dipalmitoylphosphatidylcholine (DPPC) consists of two overlapping absorptions at 1741 and 1727 cm^{-1}. The absorptions can be assigned to hydrogen-bonded and non-hydrogen-bonded C=O groups.

Perhaps the most widely studied of lipid behaviour using infrared spectroscopy is thermal phase behaviour. For example, DPPC shows three phase transitions, and it has been found that these transitions can be associated with characteristic changes in the infrared spectrum.

6.6. DRUG ANALYSIS

The GC–IR technique is an appropriate method for drug analysis as it can be used for isomer separation or contaminant detection. This technique is emerging as a complimentary method to GC–MS, which is generally accepted as the most definitive method for identification of drugs of abuse in urine, for example.

Amphetamines are one class of drug which have been succcssfully differentiated by using GC–IR. Amphetamines are structurally similar molecules which can be easily misidentified. In particular, there are some common 'designer drug' derivatives of amphetamine itself. However, although such similar compounds cannot be differentiated by their mass spectra, there are prominent differences in their infrared spectra. Figure 6.6 illustrates the FT–IR spectrum of an amphetamine drug which has been detected by using the GC–IR technique.

6.7. POLLUTION MONITORING

The routine monitoring of vapour levels of toxic chemicals in the work

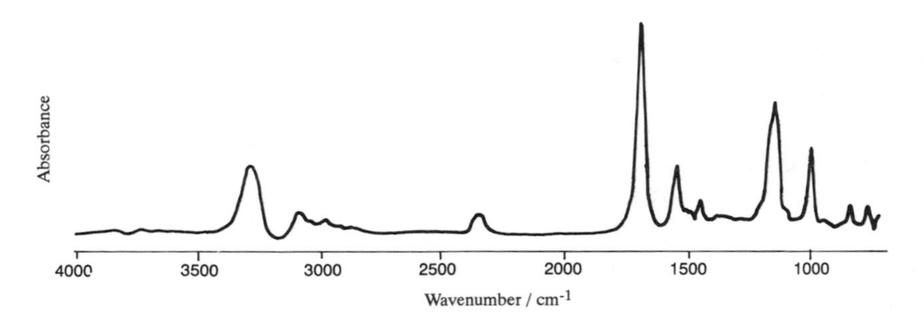

Fig. 6.6. *The GC–IR spectrum of an amphetamine*

environment has led to the development of the infrared technique for the automatic quantitative analysis of noxious chemicals in air.

The technique is based on long-pathlength cells, along with a pump to circulate the air sample through the instrument. Instruments are dedicated to a single contaminant in air, employing a filter system for wavelength selection. These are commonly used for SO_2, HCN, phosgene, NH_3, H_2S, formaldehyde and HCl. More complex instruments are tunable for a particular analytical band and usually give a direct read-out in ppm or vol% of the pollutant. The pathlength of the cell can also be varied, from less than a metre to 20 metres. Typical applications are the monitoring of formaldehyde in plastic and resin manufacture, anaesthetics in operating theatres, degreasing solvents in a wide variety of industries and carbon monoxide in garages. Table 6.6 shows some common pollutants with recommended frequencies for their detection. The minimum levels quoted are for a 20 m cell.

Table 6.6. *Common pollutants and recommended detection frequencies*

Compound	Analytical frequency (cm^{-1})	Recommended pathlength (m)	Minimum detectable concentration (ppm)
CH_3CN	1042	20	5.00
NH_3	962	20	0.20
C_6H_6	672	20	0.30
$CH_3CH_2COCH_3$	1176	20	0.15
CCl_4	794	20	0.06
CS_2	2203	20	0.50
CO_2	2353	0.75	0.50
CO	2169	20	0.20
CCl_2F_2 (Freon)	1099	0.75	0.02
CBr_2F_2	1087	2.25	0.02
SO_2	1163	20	0.50
HCN	3290	20	0.40
CH_3NH_2	2941	20	0.10
CH_2CHCl	917	20	0.30

6.8. COAL

DRIFT (see Chapter 3) is a suitable technique for evaluating the chemistry and structure of fossil fuels and can be used to measure water content, for instance. Coals are quite complex molecules, but there are some recognisable bands observed for these materials. A notable example is a highly skewed band due to hydrogen-bonded O—H stretching in the region 3600–2000 cm^{-1}. There is also a small band in the range 3100–3000 cm^{-1} which is due to C—H stretching. A very strong band is found near 1600 cm^{-1}, which is observed for conjugated polynuclear aromatic compounds. This mode overlaps with bands in the region 1450–1000 cm^{-1} that are characteristic of C—H and O—H bending modes. In addition, a triplet is observed in the range 900–700 cm^{-1}, arising from out-of-plane C—H bending modes of aromatic structures.

6.9. MINERALS

Multicomponent analysis can be readily applied to the infrared spectra of minerals. Minerals contain non-interacting components and so the spectrum of a mineral can be analysed in terms of a linear combination of the spectra of the individual components. However, the spectra of such solids exhibit a marked particle-size dependency. The particle size should be reduced (to 325 mesh) prior to preparation of an alkali halide disc. The pellet preparation involves separate grinding and dispersion steps because it is found that minerals tend not to be effectively ground in the presence of an excess of KBr.

Figure 6.9 illustrates the analysis of a mineral containing seven components. The measured sample spectrum is shown, as well as the calculated spectrum based on reference spectra of a variety of standard mineral components. The residual difference spectrum shows that the error between the two spectra is quite small.

6.10. CATALYSTS

Common applications of infrared spectroscopy in catalysis are the identification of absorbed species and studies of the manner in which

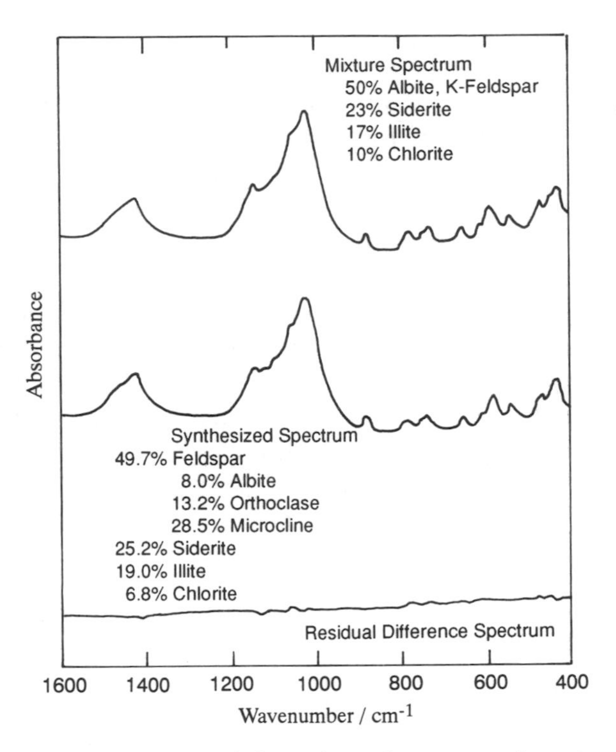

Fig. 6.9. *Analysis of the infrared spectra of a mineral*

these species are chemisorbed on the surface of a catalyst. One of the best studied systems is carbon monoxide on metals. The C—O stretching frequency provides an excellent indicator for the way that CO is bound to the substrate. Linearly absorbed CO appears in the region 2130–2000 cm^{-1}, twofold bonded CO occurs between 2000–1880 cm^{-1}, threefold bonded CO between 1880–1800 cm^{-1}, and fourfold bonded CO is observed below 1800 cm^{-1}. The exact frequency of the C—O stretching mode is dependent on the substrate metal, its surface structure and the degree of CO coverage on the surface.

Summary

This chapter began with a detailed look at the infrared spectra of a range of classes of organic compounds. Group frequencies for each class of compound were summarised.

The infrared spectra of some of the main classes of inorganic compounds, including boron, silicon and phosphorus compounds, were also described.

Appropriate sampling techniques and examples were then discussed for a range of applications of infrared spectroscopy. These include polymers, biological systems, pollution monitoring, drugs, minerals, coals and catalysts.

Objectives

On completion of this chapter you should be able to:

- recognise reliable absorption frequencies for particular functional groups;
- choose appropriate sampling techniques for a range of applications in infrared spectroscopy;
- choose appropriate methods of analysis for a range of applications in infrared spectroscopy.

7. Identification of Unknown Compounds

7.1. INTRODUCTION

To help your understanding of the subject, in this chapter we will work through some examples of the interpretation of infrared spectra of unknown compounds, trying to get as much structural information as possible. It is usually not possible by examination of the infrared spectrum of a compound alone to identify it unequivocally. It is normal to use infrared spectroscopy in conjunction with other techniques, such as chromatographic methods, mass spectrometry, NMR spectroscopy and various other spectroscopic techniques.

7.2. STRATEGY

There are a few general rules that can be stated to help you to use an infrared spectrum for the determination of a structure. However, the most effective way to learn is through practice.

The following is a suggested strategy for spectrum interpretation:

1. Look first at the high-frequency end of the spectrum ($>1500\ cm^{-1}$) and concentrate initially on the major bands.

2. For each band, short list the possibilities by using a correlation chart.

3. Use the lower-frequency end of the spectrum for the confirmation or elaboration of possible structural elements.

4. Do not expect to be able to assign every band in the spectrum.

5. Keep cross-checking wherever possible. For example, an aldehyde should absorb near 1730 cm^{-1} and also in the region 2900–2700 cm^{-1}.

6. Exploit negative evidence as well as positive evidence. For example, if there is no band in the 1850–1600 cm^{-1} region, it is most unlikely that a carbonyl group is present.

7. Band intensities should be treated with some caution. Under some circumstances they may vary considerably for the same group.

8. Take care when using small frequency changes. These can be influenced by whether the spectrum was run as a solid or as a liquid, or in solution. If run in solution, some bands are very solvent sensitive.

9. Do not forget to subtract solvent bands — these could be confused with bands from the sample

Advances in computer retrieval techniques have extended the range of information available from an infrared spectrometer by allowing comparison of an unknown spectrum with a band of known compounds. This was always possible in the past by manually searching libraries of spectra. Once the compound was classified the atlas of spectra could be searched for an identical or very similar spectrum. The advent of more efficient microcomputers has speeded up this search process. The programs work by hunting through stored data to match intensities and frequencies of absorption bands. The computer will then output the best fits and the names of the compounds found.

7.3. EXAMPLES

Example 1

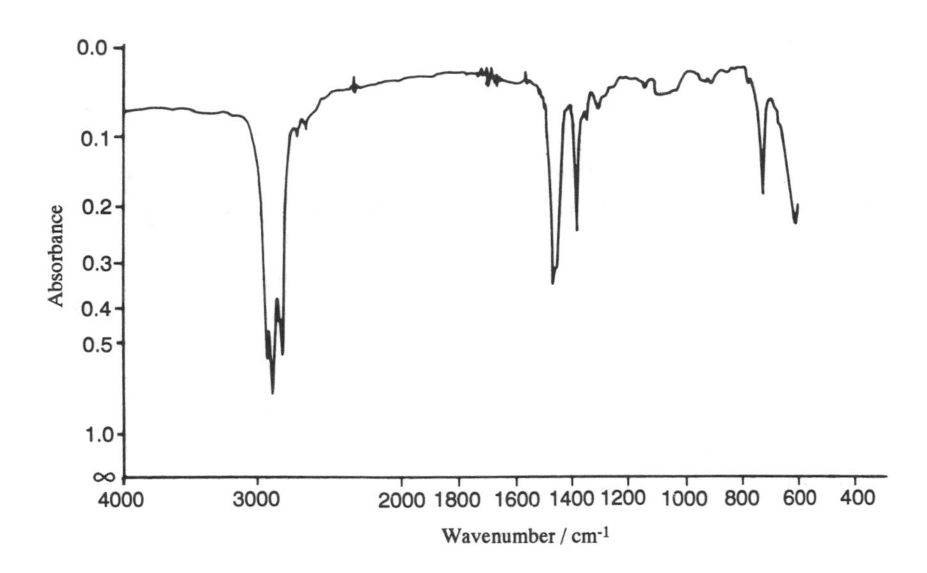

Fig. 7.3a. *The infrared spectrum of an unknown hydrocarbon ($C_{10}H_{22}$)*

Figure 7.3a shows the spectrum of a liquid compound with the molecular formula $C_{10}H_{22}$. Note the simplicity of the spectrum. The C—H stretching bands confirm that no unsaturation is present since there are no bands above 3000 cm^{-1}. The band at 1467 cm^{-1} is the scissoring frequency of CH_2 groups and that at 1378 cm^{-1} is the symmetrical bending mode of a CH_3 group. The absence of bands between 1300 and 750 cm^{-1} suggests a straight-chain structure, while the band at 782 cm^{-1} tells us that there are four or more CH_2 groups in the chain. The compound is in fact n-decane.

Example 2

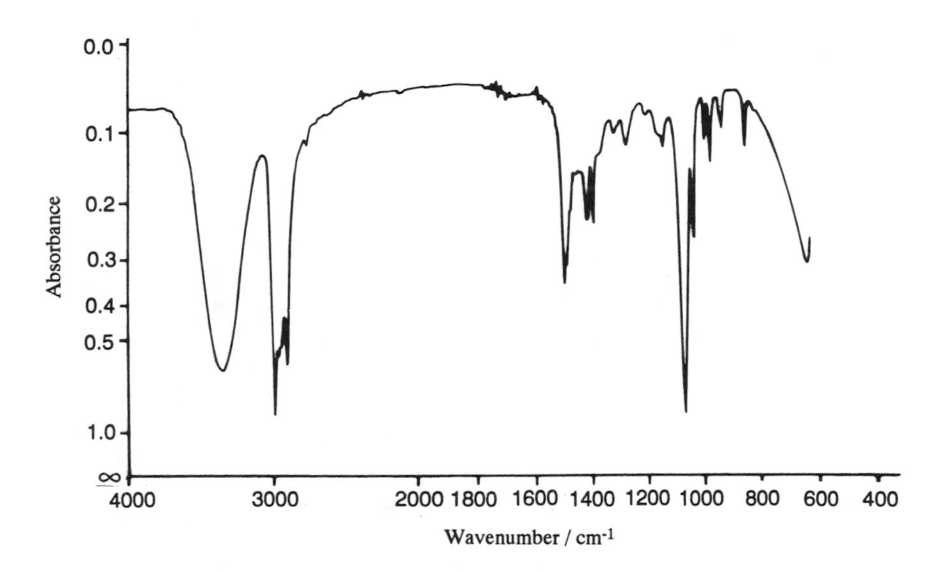

Fig. 7.3b. *The infrared spectrum of an unknown liquid ($C_4H_{10}O$)*

Figure 7.3b shows the spectrum of a saturated molecule and we can see that there are no C—H stretching absorptions above 3000 cm^{-1}. The broad band between 3700 and 3200 cm^{-1} tell us that it is an alcohol or a phenol. A C_4 alcohol fits the molecular formula. There is no band at 720 cm^{-1}, so the compound must be branched. There is also a doublet at 1386 and 1375 cm^{-1}, indicating a isopropyl group. The very strong band at 1040 cm^{-1} is due to C—O stretching. The compound must be 2-methylpropan-1-ol, $(CH_3)_2CHCH_2OH$.

Example 3

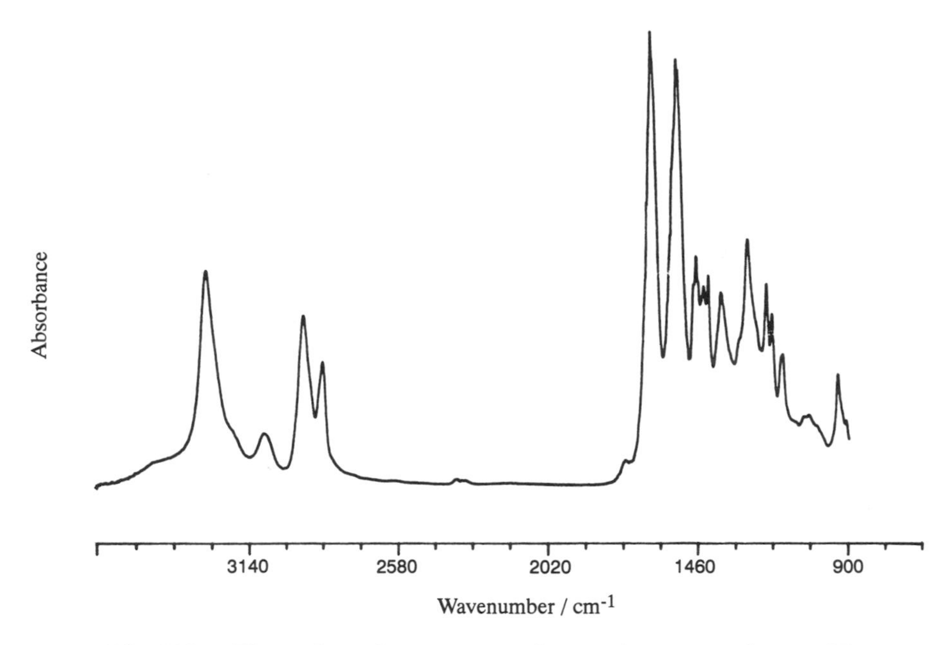

Fig. 7.3c. *The infrared spectrum of an unknown polymer film*

Figure 7.3c shows an infrared spectrum of a polymer film. On inspection of the higher-frequency end of the spectrum, we can see that there are bands at 2867 and 2937 cm^{-1}, due to symmetric and asymmetric C—H stretching, respectively. There is also a band at 3300 cm^{-1} due to N—H stretching. A band at 1640 cm^{-1} is indicative of C═O stretching. The presence of these modes suggests a polyamide and the polymer is, in fact, nylon-6. Hydrogen bonding plays a significant role in the spectra of nylons. The N—H stretching mode is due to hydrogen-bonded groups, while the broad shoulder which appears at around 3450 cm^{-1} in Figure 7.3c is due to non-hydrogen-bonded N—H bonds.

Example 4

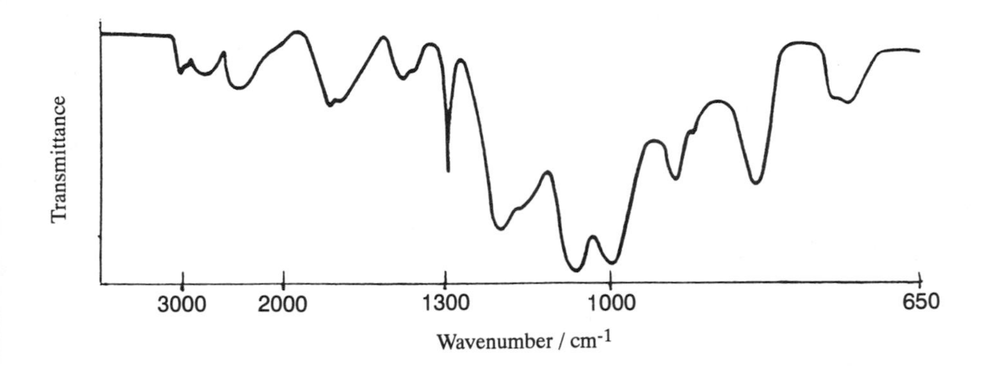

Fig. 7.3d. *The infrared spectrum of an unknown phosphorus compound ($C_2H_7PO_3$)*

Figure 7.3d shows the spectrum of a phosphorus compound with the empirical formula $C_2H_7PO_3$. The broad bands appearing at frequencies above 1500 cm^{-1} are associated with hydrogen bonding and can be attributed to a P(O)OH group. The sharper band at 1300 cm^{-1} can be assigned to P–CH_3 stretching. The broad band at 1200 cm^{-1} indicates a P=O group and confirms the presence of the P(O)OH group. In addition, the presence of a band at 1055 cm^{-1} tells us that the molecule contains a P—O—CH_3 group. The absence of sharp bands in the 2500–2000 cm^{-1} range shows that a P—H group cannot be present. Combining this information gives us a phosphoric acid with the following molecular formula:

CH_3 O
P
OCH_3 OH

7.4. EXAMPLES FOR FURTHER PRACTICE

This section contains a series of spectra which you can use to practice the information you have learnt from working through this test. You should be able to at least assign the functional groups in the molecules. Do not be too concerned if you cannot derive the molecular formula of the unknown compound. Interpretation of these spectra are given at the end of the text.

Practice Example 1

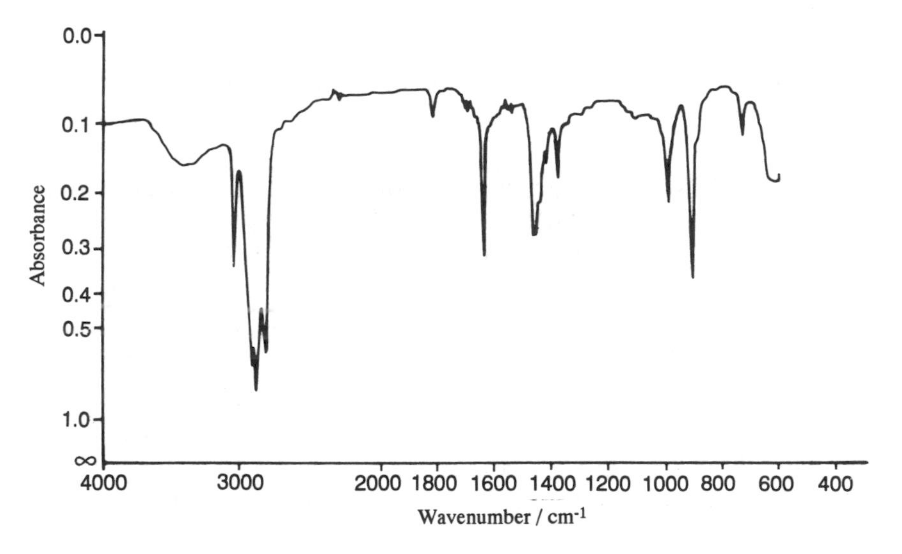

Fig. 7.4a. *The infrared spectrum of an unknown hydrocarbon (C_8H_{16})*

Practice Example 2

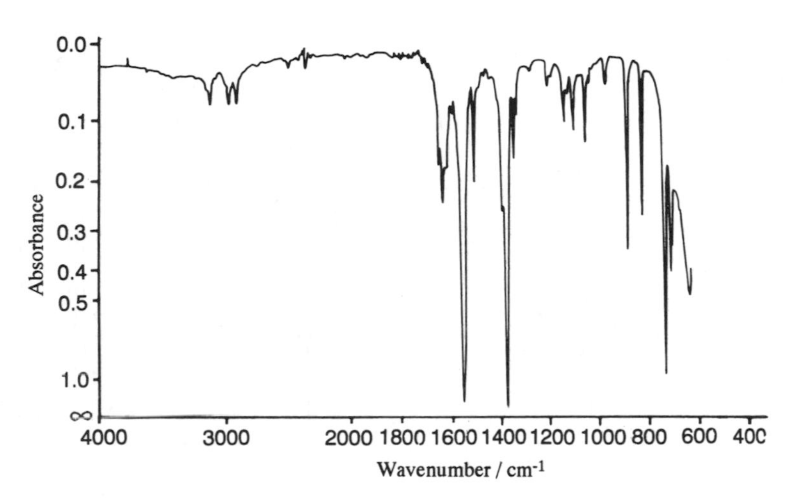

Fig. 7.4b. *The infrared spectrum of an unknown liquid*

Practice Example 3

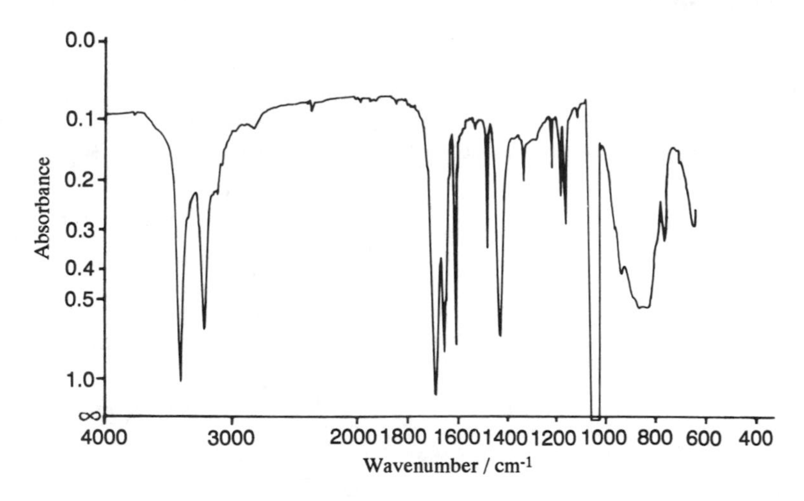

Fig. 7.4c. *The infrared spectrum of an unknown compound (C_7H_7NO) in chloroform solution*

Practice Example 4

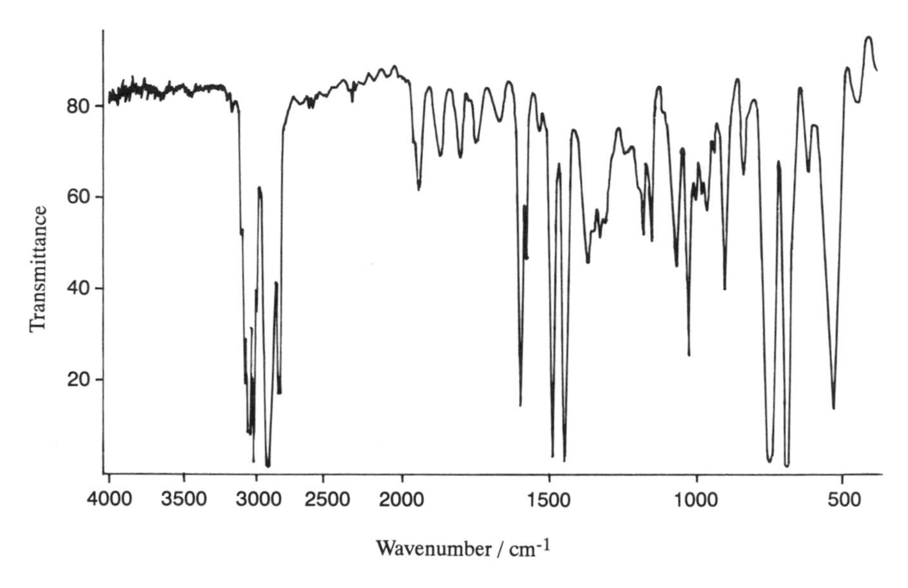

Fig. 7.4d. *The infrared spectrum of an unknown polymer film*

Practice Example 5

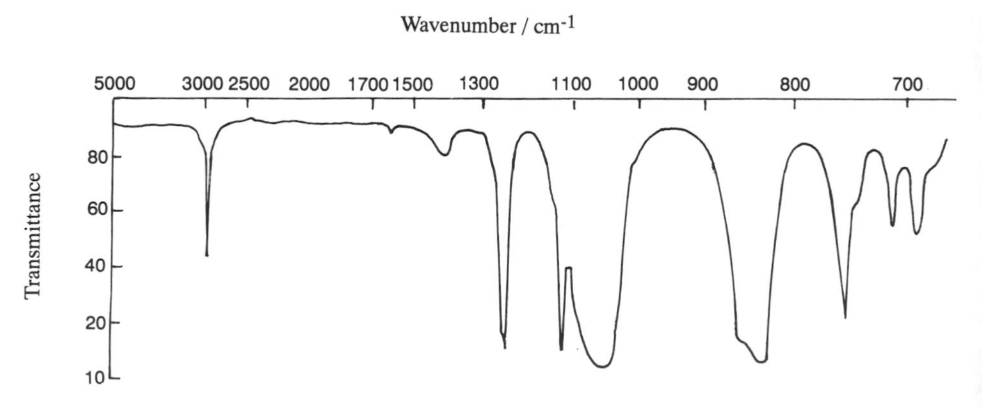

Fig. 7.4e. *The infrared spectrum of an unknown silicon-based polymer*

Summary

This chapter provided worked examples of how to interpret an infrared spectrum.

A set of revision spectra were provided for future study.

Objectives

On completion of this chapter you should be able to distinguish possible structures of molecules from the infrared spectra of unknown compounds.

Appendix

Appendix 1

CORRELATION TABLE FOR ORGANIC MOLECULES

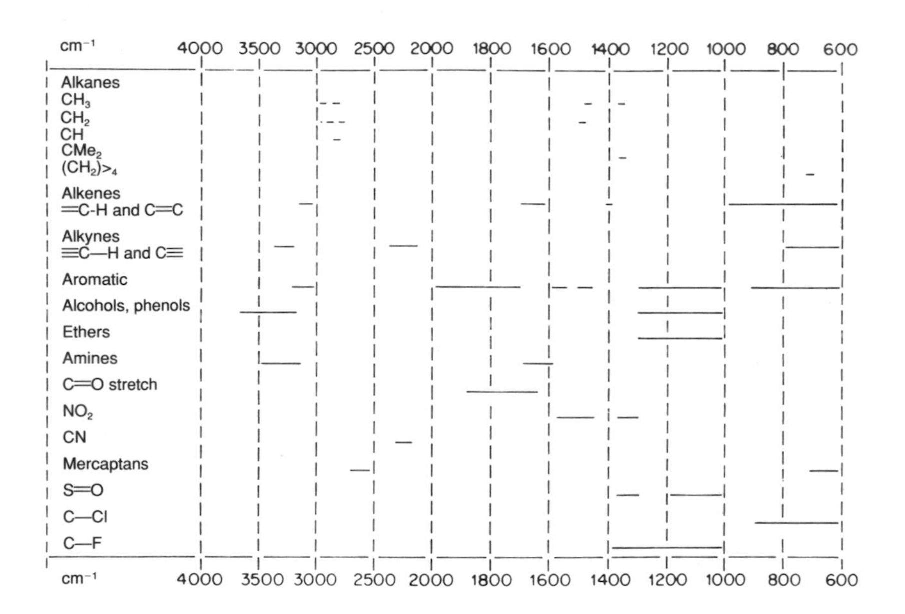

Appendix 2

CORRELATION TABLE FOR THE OUT-OF-PLANE BENDING VIBRATIONS OF BENZENOID COMPOUNDS

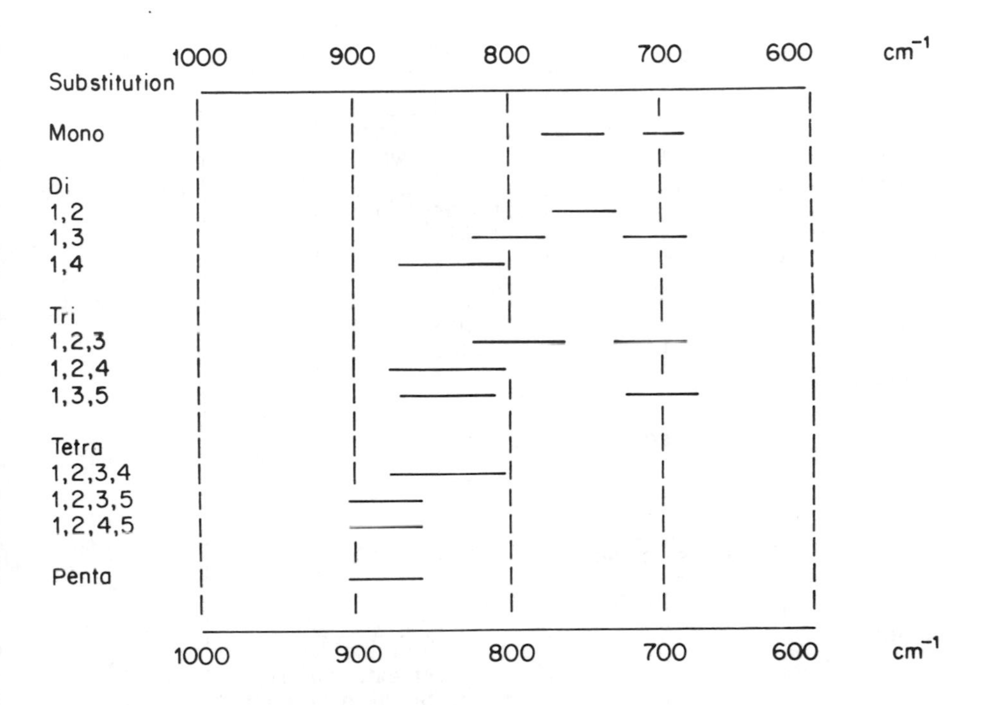

Appendix 3

CORRELATION TABLE FOR CARBONYL STRETCHING FREQUENCIES

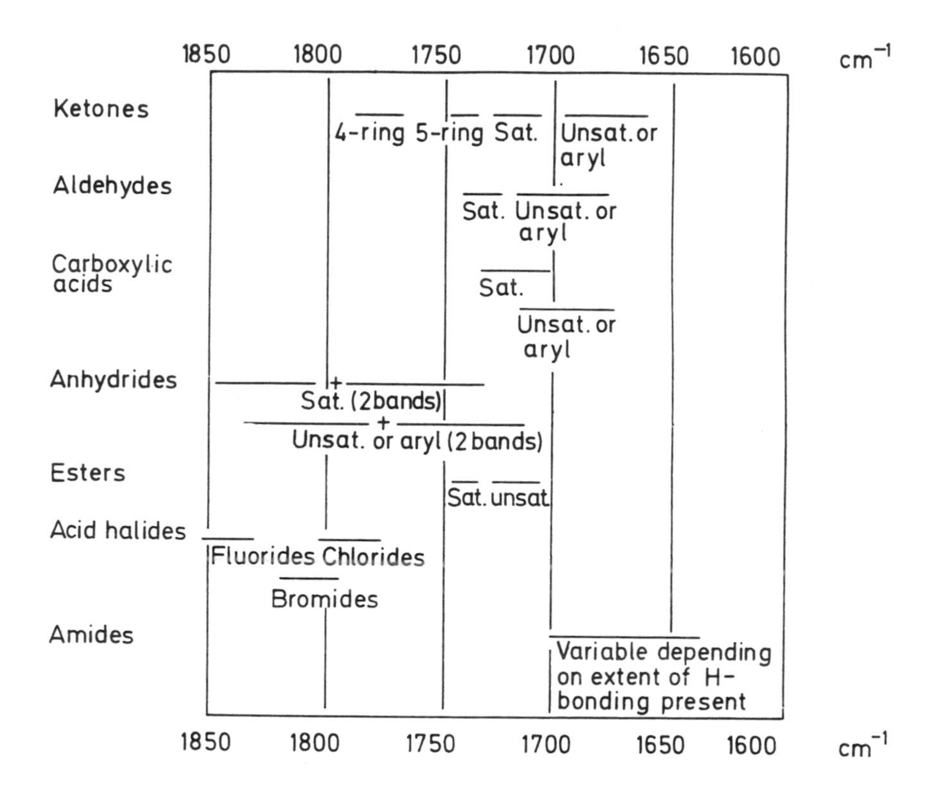

Appendix 4

CORRELATION TABLE FOR THE OUT-OF-PLANE BENDING VIBRATIONS OF ALKENES

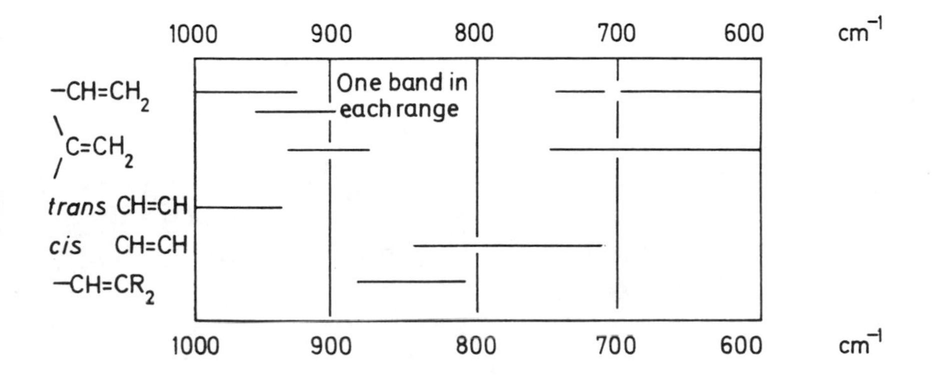

Self-assessment Questions and Responses

SAQ 1.2

The molecule hydrogen chloride (HCl) absorbs infrared radiation at 2881 cm^{-1}. Calculate the following:

(i) the wavelength of this radiation;

(ii) the frequency of this radiation;

(iii) the energy change associated with this absorption.

Response

(i) $$\lambda = \frac{1}{\bar{\nu}} = \frac{1}{2881\ \text{cm}^{-1}} = 3.471 \times 10^{-4}\ \text{cm} = 3.471\ \mu\text{m}$$

(ii) $$\nu = \frac{c}{\lambda} = \frac{2.998 \times 10^{8}\ \text{m s}^{-1}}{3.471 \times 10^{-6}\ \text{m}} = 8.637 \times 10^{13}\ \text{Hz}$$

(iii) $$\Delta E = h\nu = 6.626 \times 10^{-34}\ \text{J s} \times 8.637 \times 10^{13}\ \text{Hz} = 5.723 \times 10^{-20}\ \text{J}$$

SAQ 1.3a

How many vibrational degrees of freedom do the following molecules possess?

(i) Methane (CH_4)

(ii) Ethyne ($HC{\equiv}CH$)

Response

Decide whether the molecule is linear or non-linear, count the atoms and apply the formulae. For linear molecules there are $3N - 5$ and for non-linear molecules $3N - 6$ degrees of freedom.

(i) Methane has five atoms and is non-linear and therefore has $3 \times 5 - 6 = 9$ vibrational degrees of freedom.

(ii) Ethyne has four atoms and is a linear molecule, hence there are $3 \times 4 - 5 = 7$ degrees of freedom.

SAQ 1.3b Given that the C—H stretching vibration for chloroform occurs at $3000\,\text{cm}^{-1}$, calculate the C—D stretching frequency for deuterochloroform.

Response

Using equation (1.9):

$$\bar{\nu} = \frac{1}{2\pi c}\sqrt{\frac{k}{\mu}}$$

and assuming that k is the same for both bonds, then we need only to calculate the ratio of the reduced masses.

For C—H:

$$\mu = \frac{m_1 m_2}{m_1 + m_2}$$

$$= \frac{12 \times 1}{12 + 1} = 0.92$$

For C—D:

$$\mu = \frac{12 \times 2}{12 + 2} = 1.71$$

The vibrational frequency is proportional to the wavenumber and inversely proportional to the square root of the above values:

$$\frac{\bar{\nu}_D}{\bar{\nu}_H} = \frac{\sqrt{\mu_H}}{\sqrt{\nu_D}} = \frac{\sqrt{0.92}}{\sqrt{1.71}} = 0.73$$

So if C—H stretching for $CHCl_3$ is at 3000 cm^{-1}, C—D stretching is expected at $3000 \times 0.73 = 2190$ cm^{-1}.

SAQ 1.3c Are the following bending vibrations active or inactive?

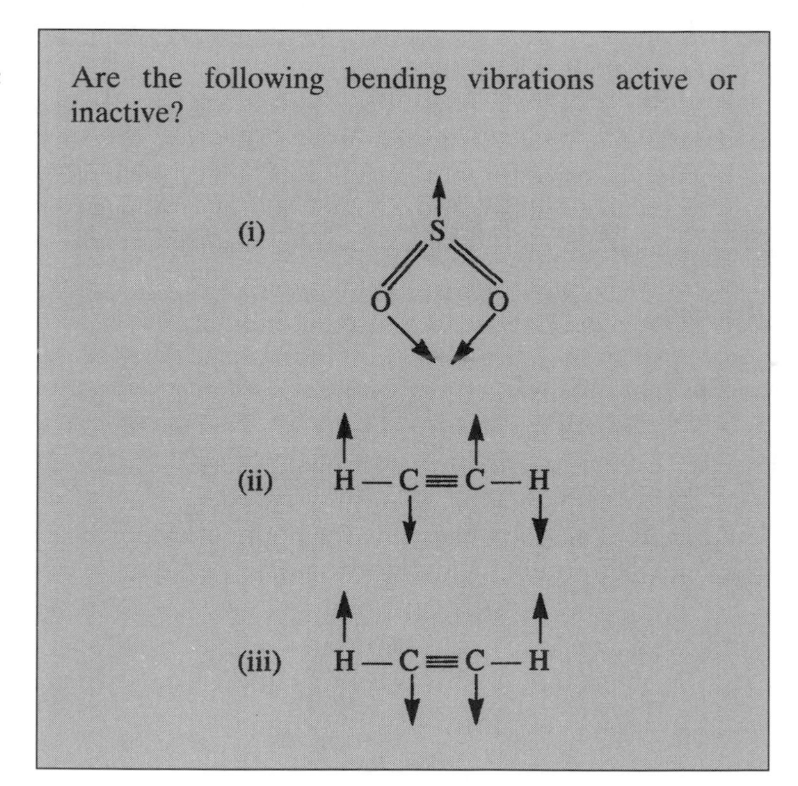

Response

(i) Active, as this vibration is exactly analogous to CO_2.

(ii) The shape of the molecule after the bending vibration will be:

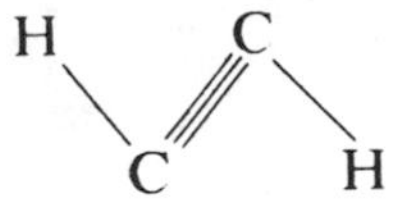

H C

C H

There was no dipole before this change and the dipoles still balance, so this is an inactive vibration.

(iii) The shape after the bending is:

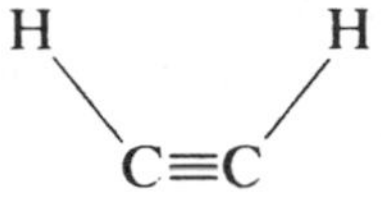

H H

C≡C

Here there is a dipole, where there was none before, so this is active.

SAQ 3.2a

What would be an appropriate material for liquid cell windows if you were wanting to examine an aqueous solution at pH 7?

Response

You will have probably noted that most of the materials listed in Table 3.2a are soluble in water. The two options available for an aqueous solution are CaF_2 or BaF_2. However, neither of these materials can be used for solutions at extreme pH values so the pH should be maintained at around a value of 7.

SAQ 3.2b

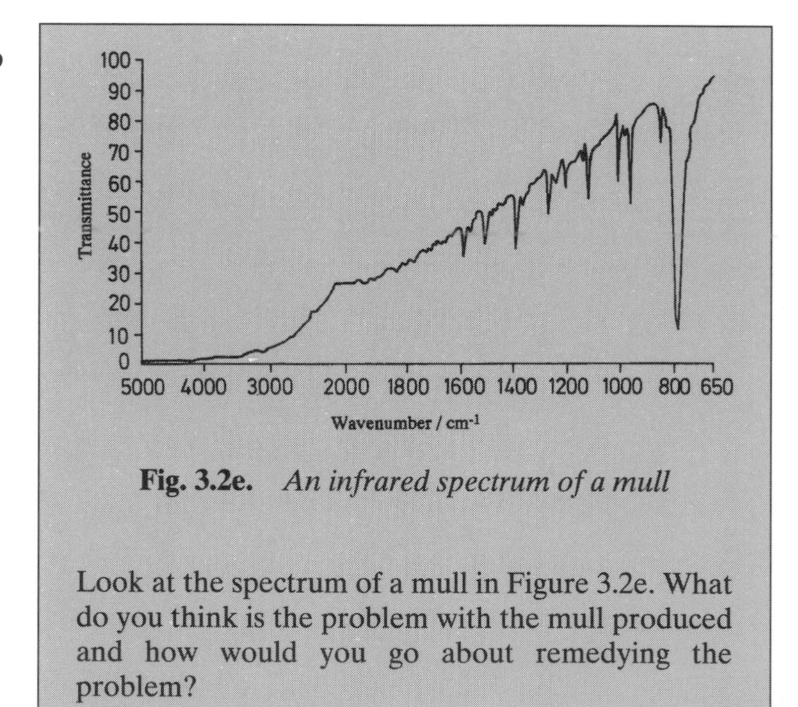

Fig. 3.2e. *An infrared spectrum of a mull*

Look at the spectrum of a mull in Figure 3.2e. What do you think is the problem with the mull produced and how would you go about remedying the problem?

Response

The crystal size of the sample is too large. This leads to scattering of radiation which gets worse at the high-frequency end of the spectrum.

Additionally, bands are distorted and consequently their positions are shifted, leading to the possibility of wrong assignments. The sample needs to be ground further with a mortar and pestle.

SAQ 3.2c

Using the interference pattern below, calculate the pathlength of the cell.

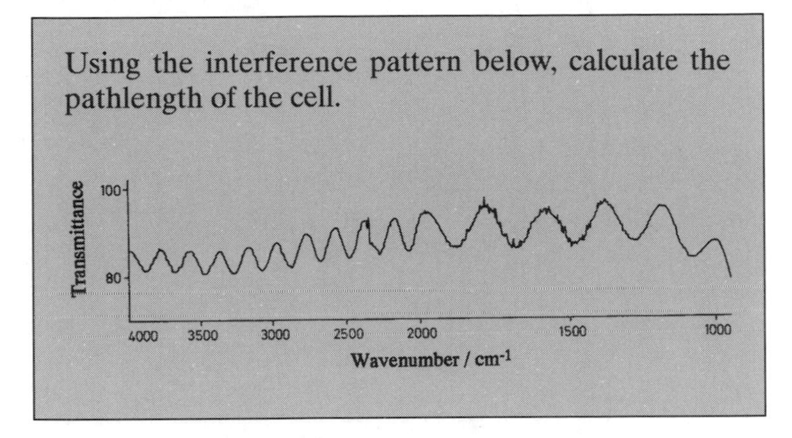

Response

You need to count the number of peak-to-peak fringes. For example, you could have chosen the region 3780–1180 cm^{-1}, since both these frequencies correspond to the tops of peaks. In this range there are 13 fringes so the pathlength of the cell, L, is calculated as follows:

$$L = \frac{n}{2(\bar{\nu}_1 - \bar{\nu}_2)}$$

$$= \frac{13}{2 \times 2600\,\text{cm}^{-1}} = 2.5 \times 10^{-3}\,\text{cm}$$

SAQ 3.3a

The spectrum of the polyamide film (refractive index 1.5) shown in Figure 3.3b was produced using an ATR cell made of KRS-5 (refractive index 2.4). If the incident radiation enters the cell crystal at an angle of 60°, what is the depth of penetration into the sample surface at:

(a) $1000\,cm^{-1}$;

(b) $1500\,cm^{-1}$;

(c) $3000\,cm^{-1}$?

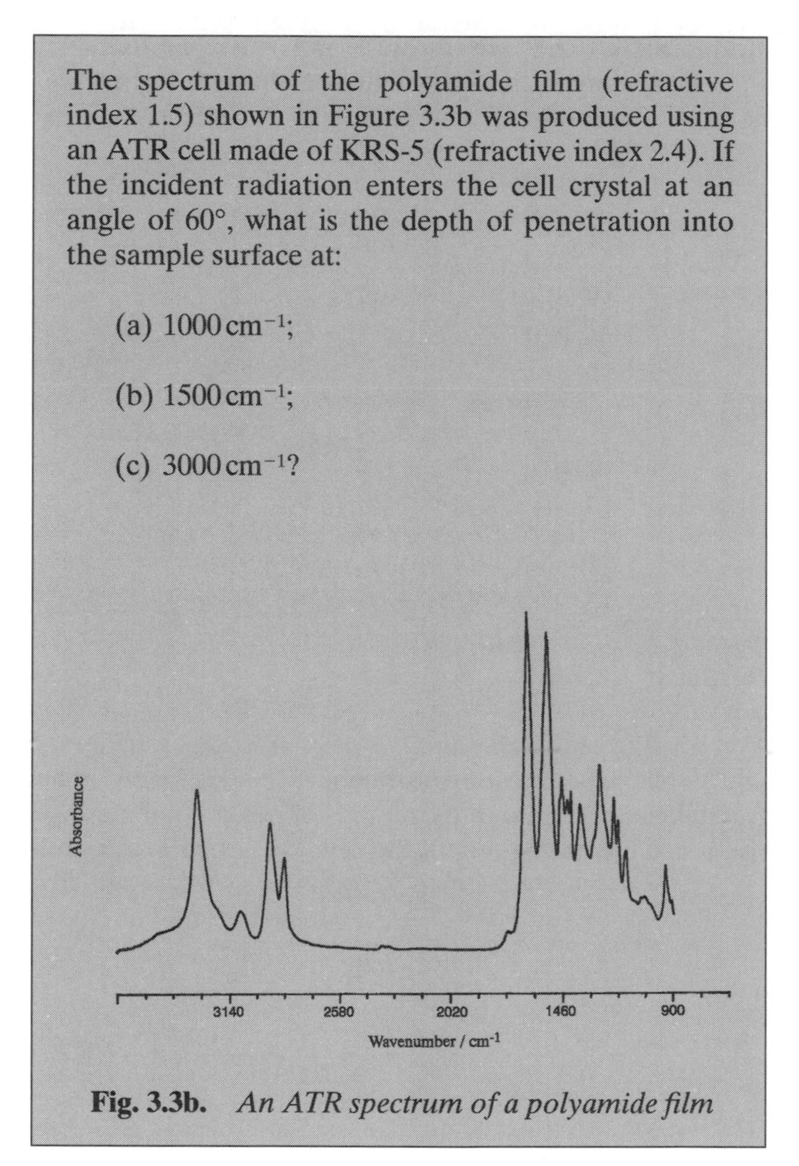

Fig. 3.3b. *An ATR spectrum of a polyamide film*

Response

(a) First, convert wavenumber into wavelength using $\lambda = 1/\nu$:

$$\lambda = \frac{1}{1000\,cm^{-1}} = 10^{-3}\,cm = 10^{-5}\,m$$

$$d_p = \frac{10^{-5}\,\text{m} \big/ 1.5}{2\pi \left[\sin 60^\circ - \left(1.5 \big/ 2.4 \right)^2 \right]^{0.5}} = 1.5 \times 10^{-6}\,\text{m} = 1.5\,\mu\text{m}$$

(b) $d_p = 1.0\,\mu\text{m}$
(c) $d_p = 0.5\,\mu\text{m}$

The depth of penetration at higher wavenumbers ($3000\,\text{cm}^{-1}$) is notably less than at lower wavenumbers ($1000\,\text{cm}^{-1}$).

SAQ 4.2a

Examine the spectra in Figures 4.2a–4.2e and classify them as below:

(i) aliphatic C—H bonds only;

(ii) aliphatic and aromatic C—H bonds;

(iii) an alkene or aromatic compound containing no aliphatic C—H bonds;

(iv) an alkyne;

(v) a deuterated compound.

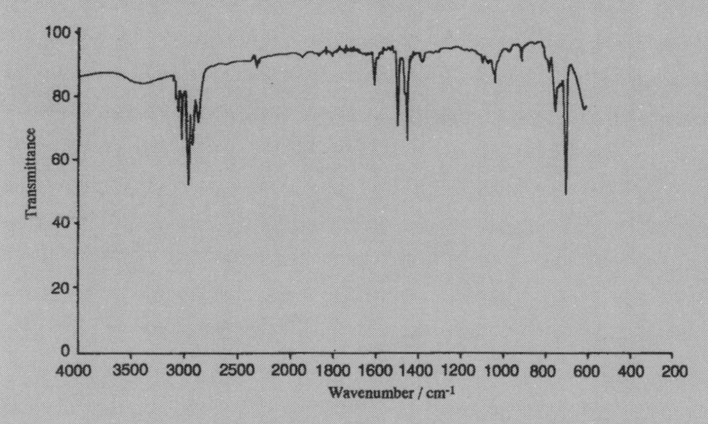

Fig. 4.2a. *Infrared spectrum of an unknown substance*

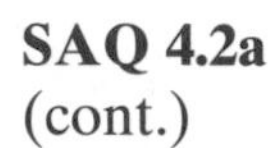

SAQ 4.2a
(cont.)

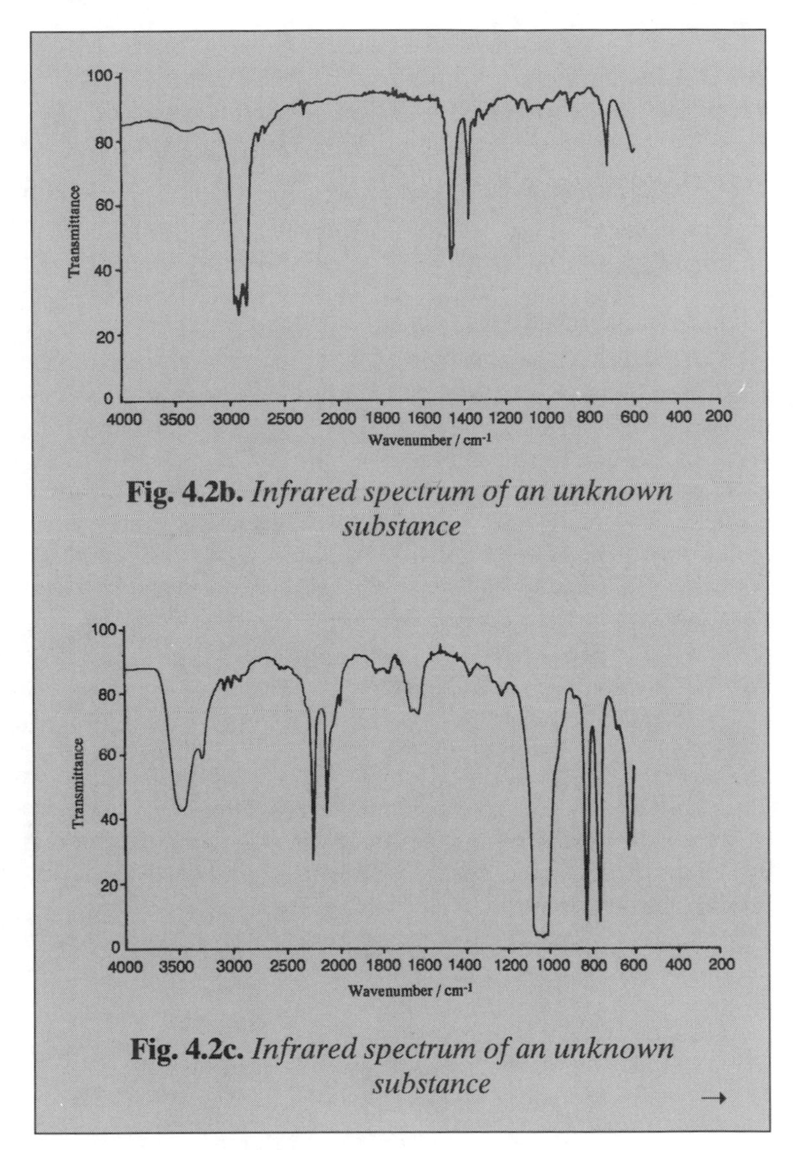

Fig. 4.2b. *Infrared spectrum of an unknown substance*

Fig. 4.2c. *Infrared spectrum of an unknown substance*

→

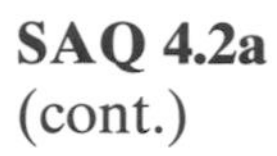

SAQ 4.2a
(cont.)

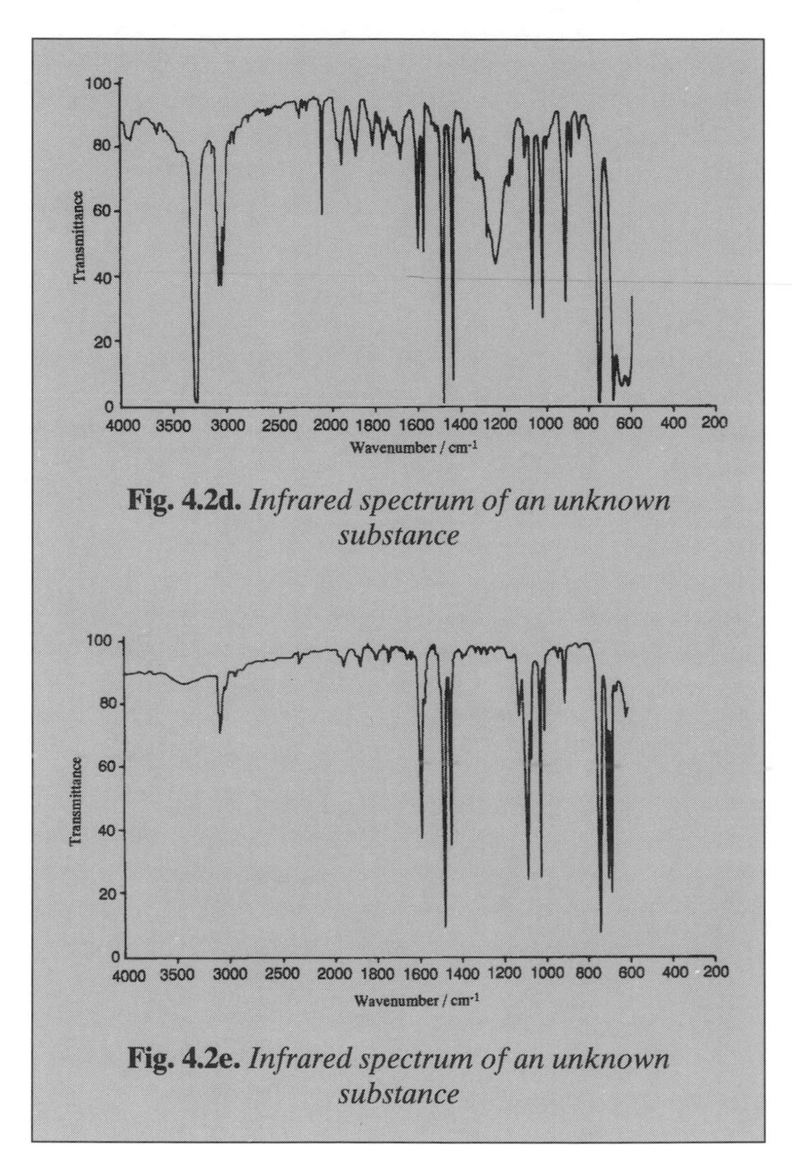

Fig. 4.2d. *Infrared spectrum of an unknown substance*

Fig. 4.2e. *Infrared spectrum of an unknown substance*

Response

Figure 4.2a contains a series of C—H stretching bands both above and below 3000 cm^{-1}. This is therefore the spectrum of a compound containing both aliphatic and aromatic C—H bonds.

Figure 4.2b has no absorptions above 3000 cm^{-1}, so this compound contains aliphatic C–H bonds only.

Figure 4.2c has only a weak absorption band around 3000 cm^{-1}, but has C—D stretching bands at 2120 and 2250 cm^{-1}, so this is a deuterated compound.

Figure 4.2d has aromatic C—H stretching bands in the region 3100–3000 cm^{-1}, but also has a strong band at 3300 cm^{-1}, indicative of the C—H stretching frequency of an alkyne.

Figure 4.2e has no aliphatic absorptions below 3000 cm^{-1} and so this is either an aromatic compound containing no aliphatic hydrogens or a simple alkene.

SAQ 4.2b

Carbon monoxide absorbs at 2143 cm^{-1}. What does this tell you about the bond order in this molecule?

Response

This is close to the C≡C triple bond absorption frequency, indicating a bond order of three for carbon monoxide.

SAQ 4.2c

The spectra in Figures 4.2g–4.2i are of 1,2-dimethylbenzene, 1,3-dimethylbenzene and 1,4-dimethylbenzene, in the region 2000–1650 cm^{-1}. Which is which?

→

SAQ 4.2c
(cont.)

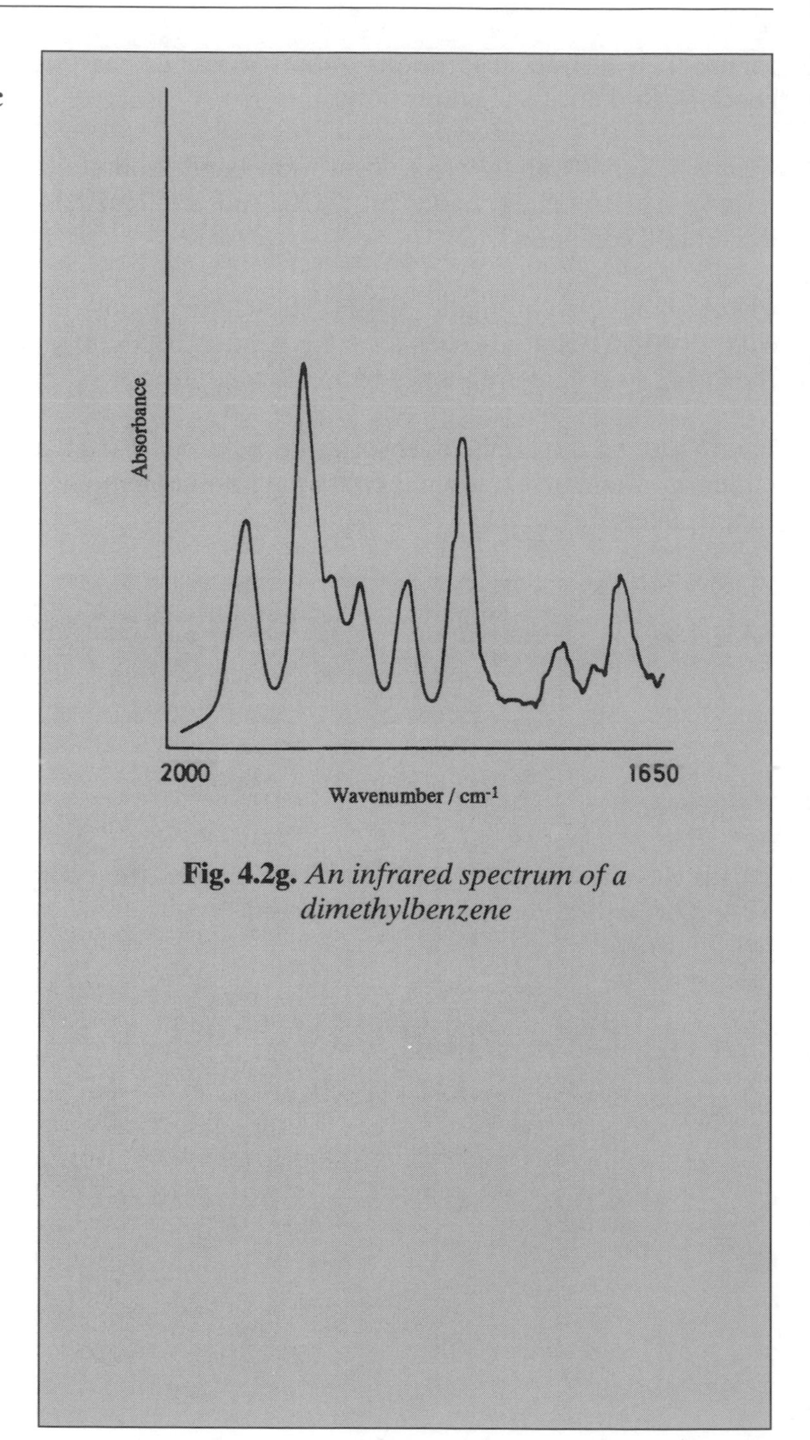

Fig. 4.2g. *An infrared spectrum of a dimethylbenzene*

SAQ 4.2c
(cont.)

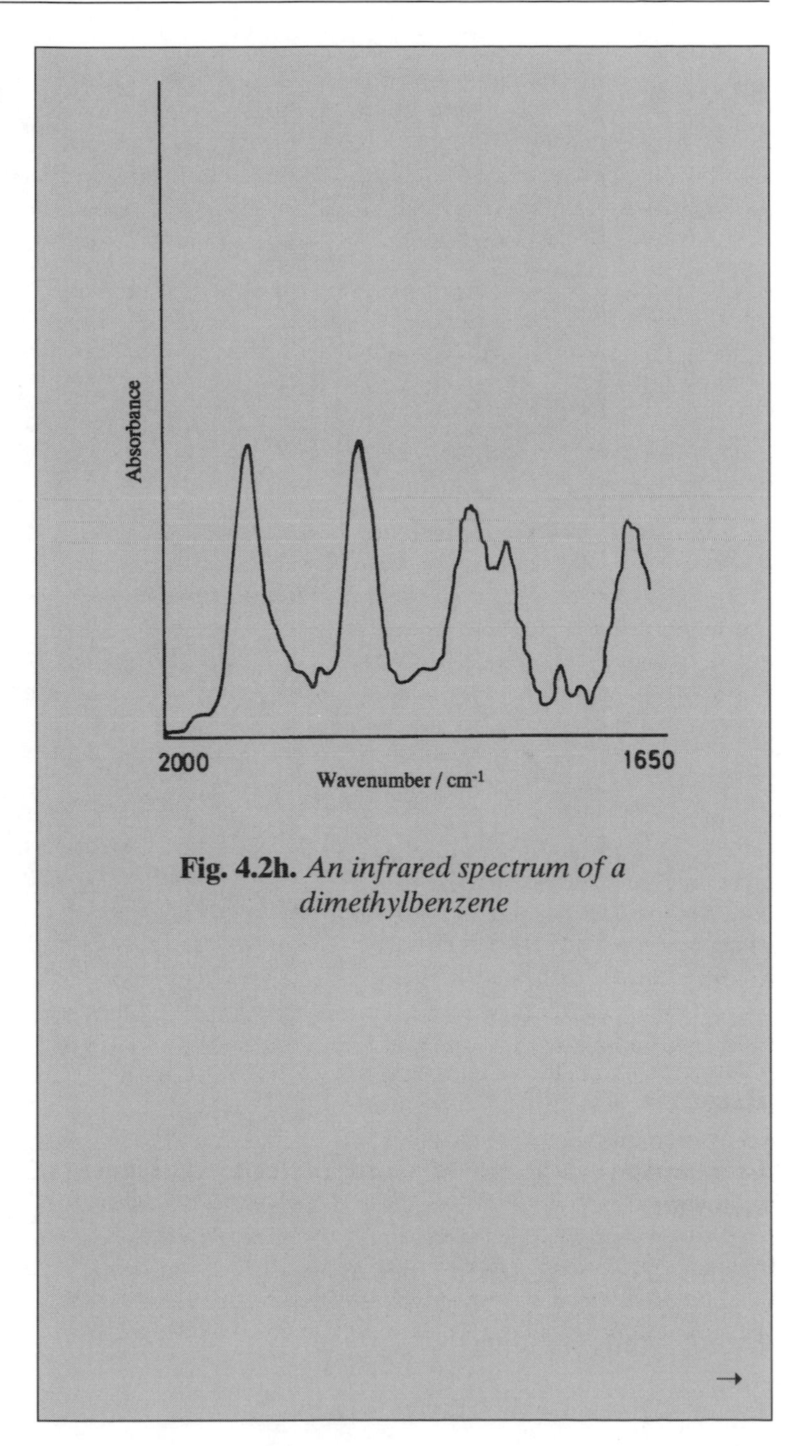

Fig. 4.2h. *An infrared spectrum of a dimethylbenzene*

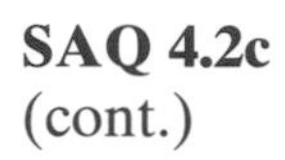

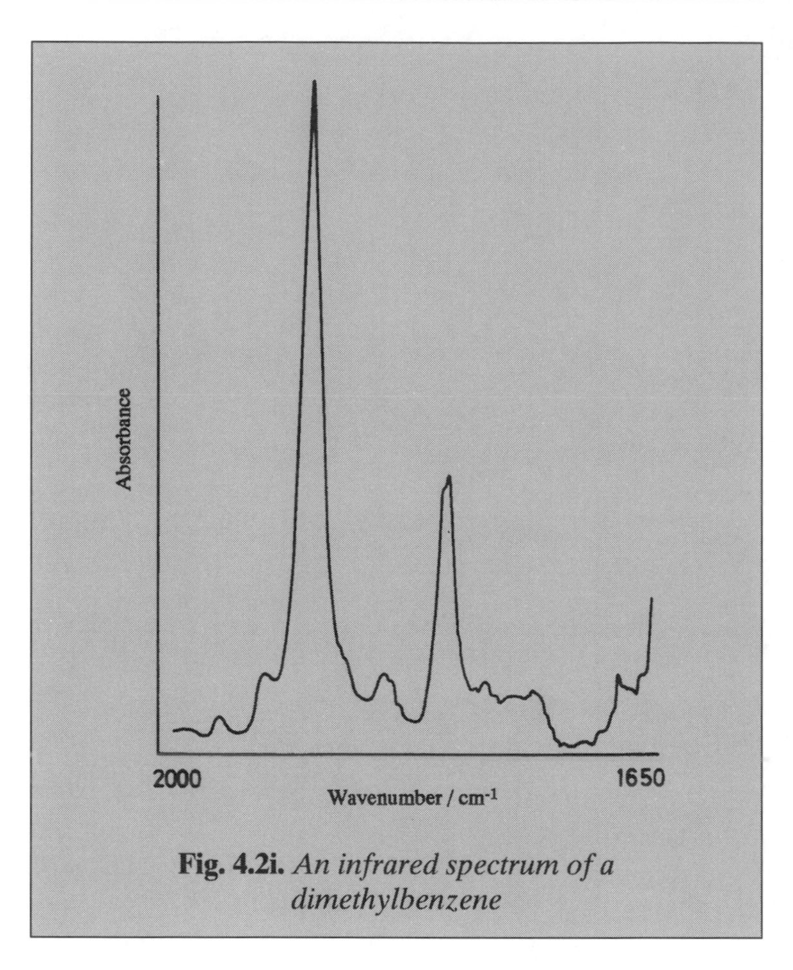

Fig. 4.2i. *An infrared spectrum of a dimethylbenzene*

Response

Comparison with the standard patterns in Figure 4.2f gives the following:

Figure 4.2g 1,2-dimethylbenzene;

Figure 4.2h 1,3-dimethylbenzene;

Figure 4.2i 1,4-dimethylbenzene.

SAQ 4.2d

Would you expect a C—O stretching mode to be more or less intense than a C—C stretching mode?

Response

The intensity of a band in the infrared region depends on the change in dipole moment during the vibration. The change for C—O stretching will be much greater than for C—C stretching, so therefore you would expect an intense absorption for the C—O stretching mode.

SAQ 4.3a

A molecule has strong fundamental bands at the following frequencies:

C—H bending at $730\,cm^{-1}$;

C—C stretching at $1400\,cm^{-1}$;

C—H stretching at $2950\,cm^{-1}$.

Write down the frequencies of the possible combination bands and the first overtones.

Response

The first overtone will occur at double the wavenumber of the fundamentals, so we would expect bands at 1460, 1800 and $5900\,cm^{-1}$. The possible combinations are at:

$$730 + 1400 = 2130\,cm^{-1};$$

$$730 + 2950 = 3680\,cm^{-1};$$

$$1400 + 2950 = 4350\,cm^{-1}.$$

SAQ 4.3b

Tetrachloroethane is expected to show only four infrared-active fundamentals. Three of these fundamentals absorb at 217, 313 and 459 cm^{-1}. The fourth is expected to occur in the region 700–800 cm^{-1}. The spectrum has two bands in this frequency range at 762 and 791 cm^{-1}. Can you account for this observation?

Response

A combination band is possible at 459 + 313 = 772 cm^{-1}. The fourth fundamental could undergo Fermi resonance with this band. This explains the two bands, almost symmetrical, about this frequency.

SAQ 4.3c

Examine the infrared spectra of ethanol in Figures 4.3d and 4.3e. Figure 4.3d is of a 10 vol% solution of ethanol in CCl_4, run at a pathlength of 0.1 mm, while Figure 4.3e is a 1 vol% solution of the same compound in the same solvent at a pathlength of 1.0 mm. Can you account for any observed differences?

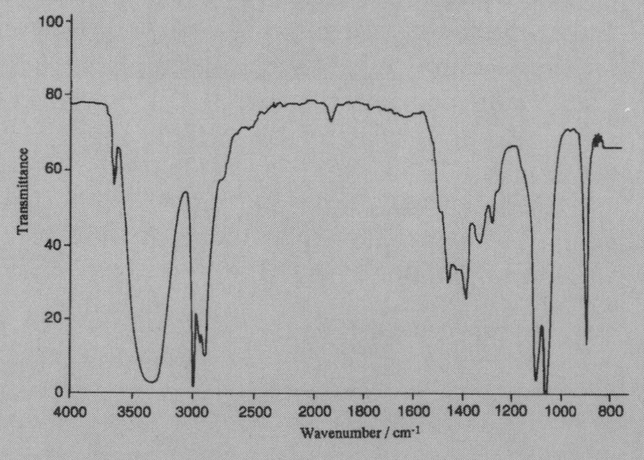

Fig. 4.3d. *The infrared spectrum of ethanol (10 vol% in CCl_4, 0.1 mm pathlength cell)*

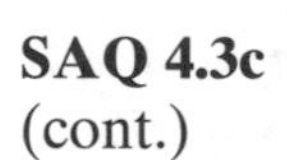

SAQ 4.3c
(cont.)

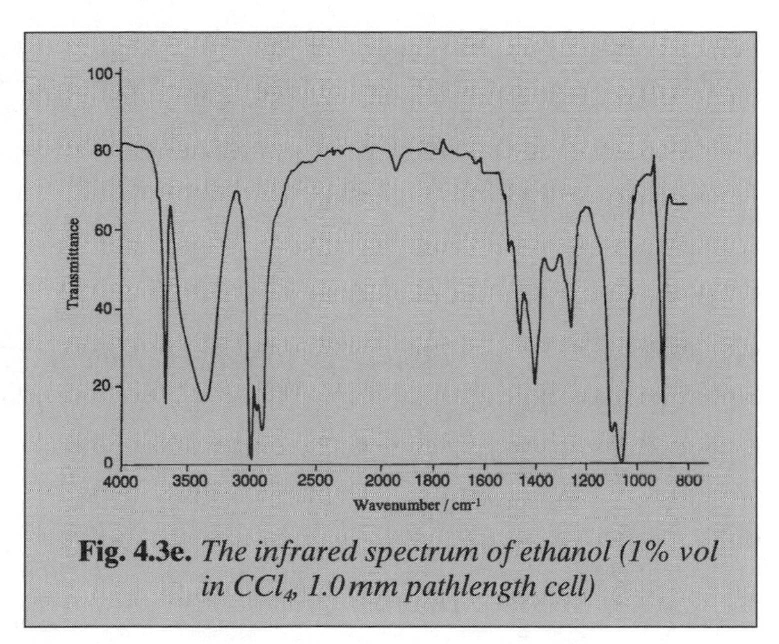

Fig. 4.3e. *The infrared spectrum of ethanol (1% vol in CCl_4, 1.0 mm pathlength cell)*

Response

The main differences between the spectra occur in the region above 3000 cm^{-1}. The 10 vol% solution has a weak, sharp absorption at 3640 cm^{-1} and a strong broad absorption centred at 3340 cm^{-1}. The 1 vol% solution shows these same two peaks, but the sharp peak becomes more intense on dilution, relative to the broad absorption. We can assign the higher-frequency sharp band to non-hydrogen-bonded molecules, while the lower-frequency broad band is assigned to hydrogen-bonded molecules. Thus, comparison of the spectra shows that the degree of hydrogen bonding in ethanol is greater in the more concentrated solution.

SAQ 5.3

A 1.0% w/v solution of hexan-1-ol has an absorbance of 0.37 at 3660 cm^{-1} in a 1.0 mm cell. Calculate its molar absorptivity at this frequency.

Response

1.0% w/v means 1.0 g dissolved in 100 cm^3 (10 g dm^{-3}). The relative molecular mass of hexan-1-ol ($C_6H_{13}OH$) is 92, so the concentration is:

$$(10/92)\ mol\ dm^{-3} = 0.11\ mol\ dm^{-3}.$$

From the Beer–Lambert Law:

$$\epsilon = A/cl = 0.37/(0.11\ mol\ dm^{-3} \times 1.0\ mm);$$

$$\epsilon = 3.4\ m^2\ mol^{-1}.$$

SAQ 5.4a

Draw a plot of absorbance against concentration from the data given below and calculate the molar absorptivity in units of $m^2\ mol^{-1}$.

Acetone in CCl_4 (vol %)	Absorbance at 1719 cm^{-1}
0.25	0.183
0.50	0.315
1.00	0.570
1.50	0.796
2.00	1.000

The absorbance values that are given were read straight from the spectrum. The baseline had an absorbance of 0.06 at 1719 cm^{-1}. The density of acetone is 0.790 g cm^{-3} and the pathlength was 0.1 mm.

Response

You were told that the baseline was at an absorbance of 0.06. We must subtract this from the absorbance values in the above table to give the actual absorbance of the peak in each spectrum.

Acetone in CCl_4 (vol%)	Corrected absorbance at 1719 cm^{-1}
0.25	0.123
0.50	0.225
1.00	0.510
1.50	0.736
2.00	0.940

The plot of absorbance versus concentration is shown in Figure 5.4d. The slope of this plot is 5.10 per vol%.

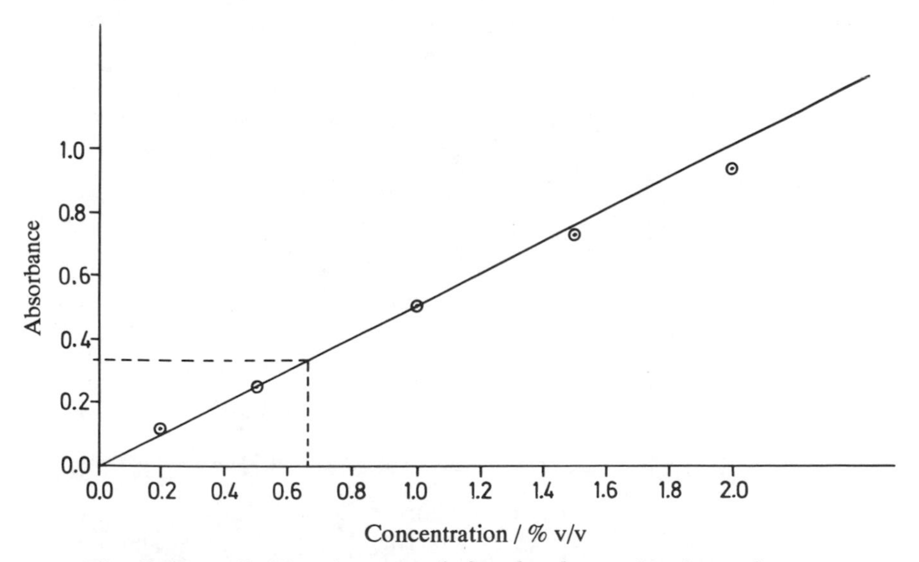

Fig. 5.4d. *Calibration graph for the determination of acetone*

We need to convert concentration units to mol dm^{-3} in order to obtain the conventional units of $m^2 mol^{-1}$ for molar absorptivity. Now, 1 vol% means we have 1 cm^3 in 100 cm^3, or 10 cm^3 in 1 dm^3. A volume of acetone of 10 cm^3 has a mass:

$$m = \rho V = 0.790\,\text{g cm}^{-3} \times 10\,\text{cm}^3 = 7.90\,\text{g}$$

The relative molar mass of acetone is 58, so the equivalent concentration is (7.90/58) mol dm^{-3} = 0.136 mol dm^{-3}.

Thus:

$$1\ \text{vol\%} = 0.136\,\text{mol dm}^{-3};$$

$$\text{slope} = 5.10/\text{vol\%} = 5.10/0.136\,\text{mol dm}^{-3} = 37.5\,\text{dm}^3\,\text{mol}^{-1};$$

$$\epsilon = \text{slope}/l = 37.5\,\text{dm}^3\,\text{mol}^{-1}/0.1\,\text{mm} = 375\,\text{m}^2\,\text{mol}^{-1}.$$

SAQ 5.4b

The infrared spectrum of a 10 vol% solution of commercial propan-2-ol in CCl_4 in a 0.1 mm pathlength cell is shown in Figure 5.4c.

(i) Determine the concentration (in mol dm^{-3}) of acetone in this solution by using the calibration curve plotted in SAQ 5.4a.

(ii) Calculate the % acetone in the propan-2-ol.

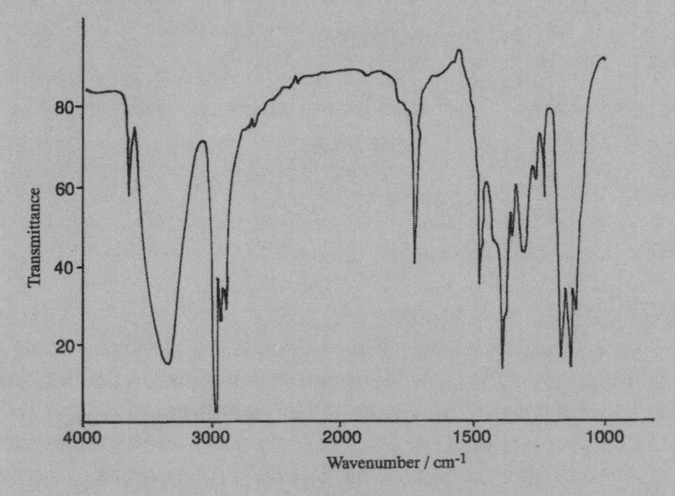

Fig. 5.4c. *An infrared spectrum of commercial propan-2-ol in carbon tetrachloride*

Response

The spectrum shows a C=O stretching band with a transmittance of 40.5%. The baseline has a transmittance of 88%. This corresponds to an absorbance value of 0.393 and gives a corrected absorbance of (0.393 − 0.056) = 0.337. From the calibration graph this corresponds to 0.67 vol% or 0.09 mol dm^{-3}. This is a 10 vol% solution in CCl_4 here so we must multiply by 10, thus giving a concentration of acetone in propan-2-ol of 0.90 mol dm^{-3} or 6.7 vol%.

SAQ 6.2a

Examine the spectrum of nonane in Figure 6.2a and describe the vibrations corresponding to the absorptions marked A, B and C.

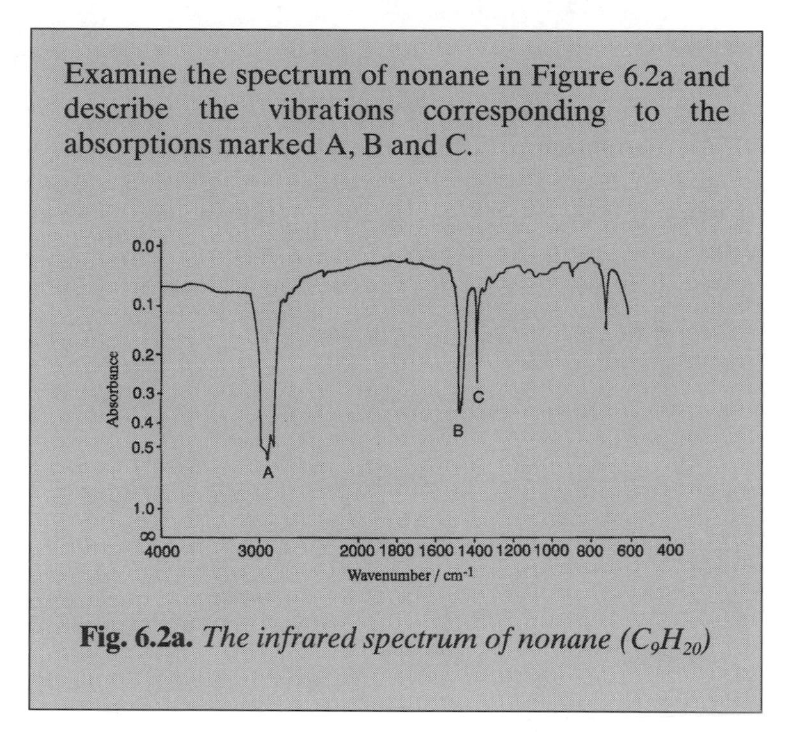

Fig. 6.2a. *The infrared spectrum of nonane (C_9H_{20})*

Response

The absorption marked A shows four bands at around 2960, 2930, 2870 and 2850 cm^{-1}.

Those at 2960 and 2870 cm^{-1} are the asymmetric and symmetric stretching absorptions of the methyl group, respectively. The others are the corresponding bands from the methylene group.

The absorption marked B shows two bands at 1465 and 1450 cm^{-1}. The higher one is the CH_2 deformation while the lower one is the asymmetric CH_3 deformation.

The band marked C is a singlet and absorbs at 1380 cm^{-1}. This is the symmetrical CH_3 deformation.

SAQ 6.2b

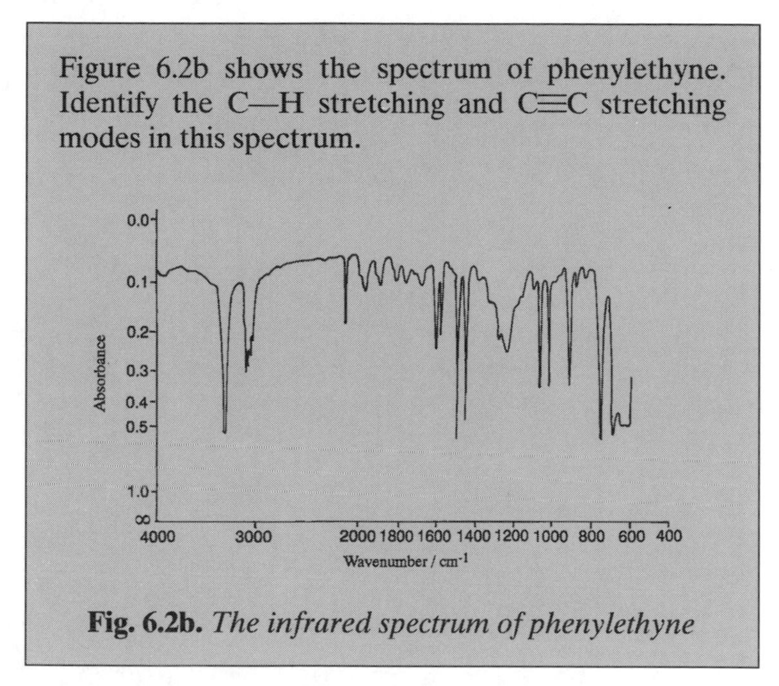

Figure 6.2b shows the spectrum of phenylethyne. Identify the C—H stretching and C≡C stretching modes in this spectrum.

Fig. 6.2b. *The infrared spectrum of phenylethyne*

Response

The stretching frequency of the hydrogen attached to a triple-bonded carbon occurs at a high frequency of 3315 cm^{-1}. This mode is sharp and fairly intense and is usually easy to assign. The C≡C stretching mode is fairly weak and is observed for phenylethyne at 2105 cm^{-1}.

SAQ 6.2c

Examine the three spectra in Figures 6.2d–6.2f, which have been obtained for various isomeric disubstituted benzenes. Which is 1,2- which 1,3- and which is 1,4-disubstituted?

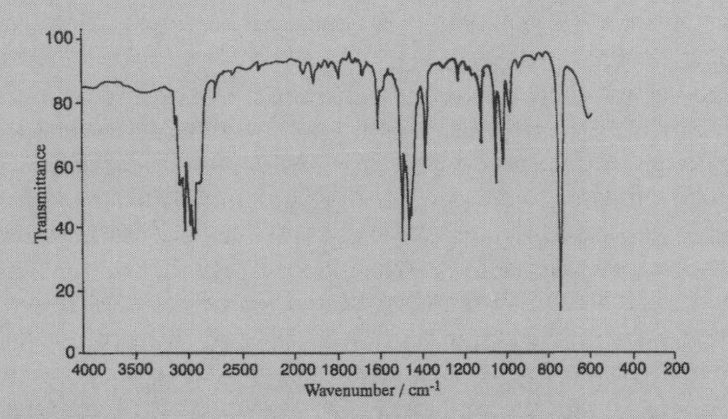

Fig. 6.2d. *The infrared spectrum of a disubstituted benzene: compound A*

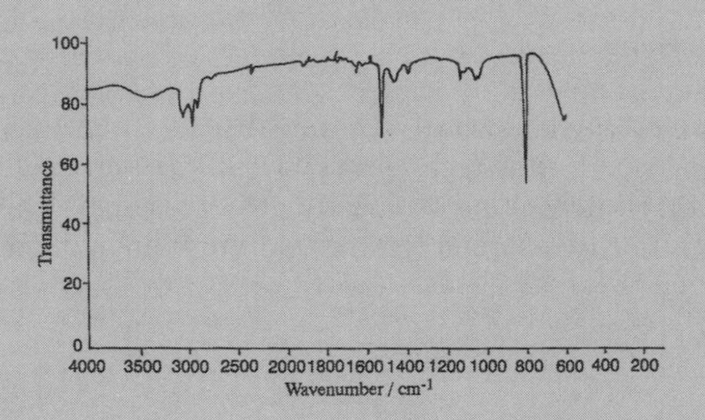

Fig. 6.2e. *The infrared spectrum of a disubstituted benzene: compound B*

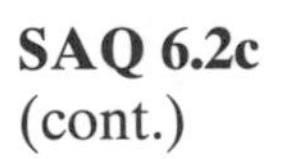

SAQ 6.2c
(cont.)

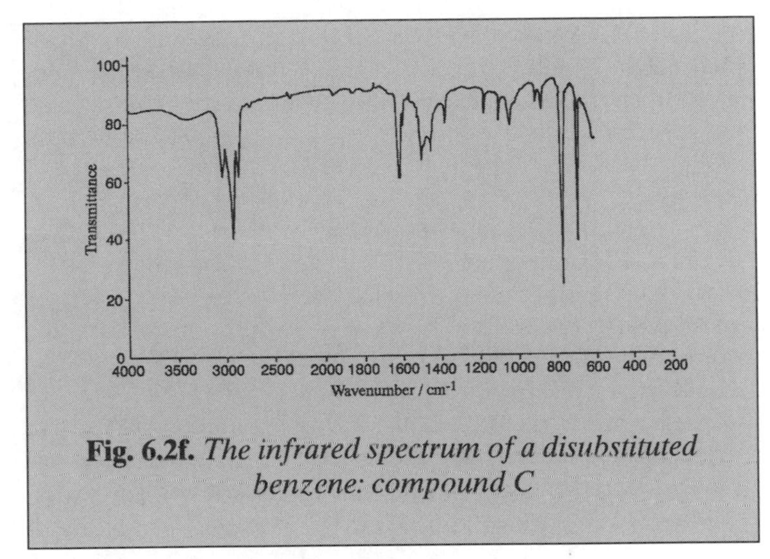

Fig. 6.2f. *The infrared spectrum of a disubstituted benzene: compound C*

Response

These compounds are very easily differentiated by using the C—H deformation bands between 600 and 1000 cm^{-1}. Compound A shows one band at 734 cm^{-1}. Compound B shows one band at 787 cm^{-1}. Compound C shows two bands at 680 and 760 cm^{-1}. Use of Figure 6.2c tells us that the only disubstitution pattern giving two peaks in this region is 1,3. It also tells us that 1,2 and 1,4 give one peak, with 1,2 absorbing at a lower frequency than 1,4. Hence A is 1,2-disubstituted, B is 1,4-disubstituted and C is 1,3-disubstituted.

SAQ 6.3

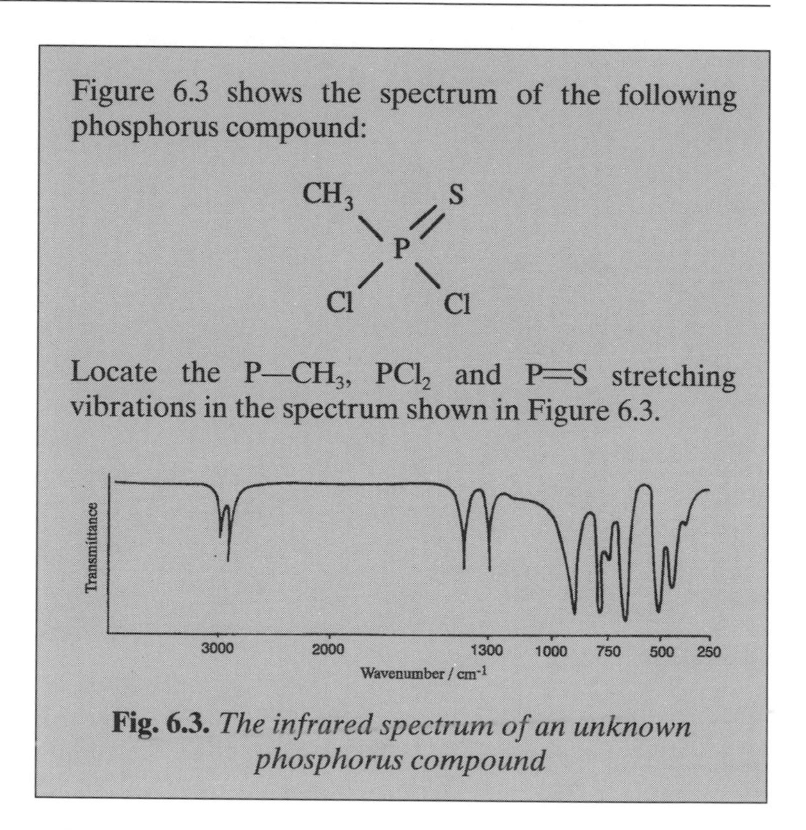

Figure 6.3 shows the spectrum of the following phosphorus compound:

Locate the P—CH_3, PCl_2 and P=S stretching vibrations in the spectrum shown in Figure 6.3.

Fig. 6.3. *The infrared spectrum of an unknown phosphorus compound*

Response

P—CH_3 stretching, 1290 cm^{-1};

P=S stretching, 660 and 780 cm^{-1};

symmetric PCl_2 stretching, 445 cm^{-1};

asymmetric PCl_2 stretching, 500 cm^{-1}.

SAQ 6.4

The polymer poly(methyl methacrylate) (PMMA) has the following structural repeat unit:

$$\left[-CH_2-\overset{\displaystyle H}{\underset{\displaystyle CO_2H}{\overset{|}{\underset{|}{C}}}}- \right]_n$$

The infrared spectrum of PMMA is shown in Figure 6.4c. Can you identify the C—C—O, C—H and C=O stretching, and OCH_3 bending modes in this spectrum?

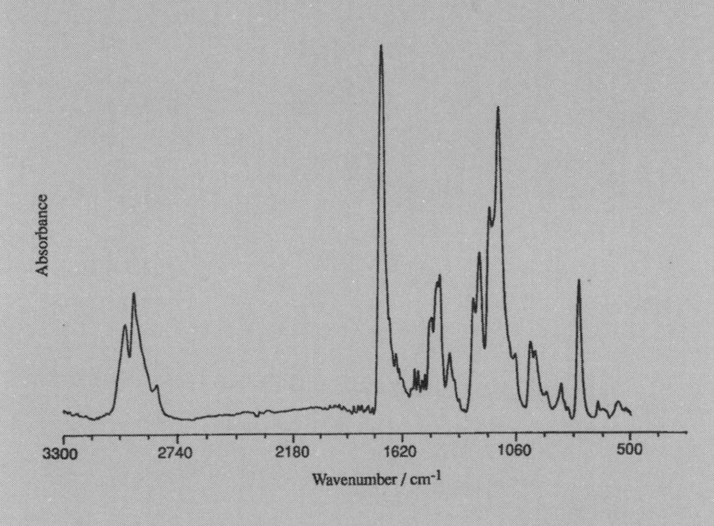

Fig. 6.4c. *The infrared spectrum of poly(methyl methacrylate)*

Response

C—C—O stretching, 1265 and 1238 cm^{-1};

C—H stretching, 2992 and 2948 cm^{-1};

C═O stretching, 1730 cm^{-1};

OCH_3 bending, 1450 and 1434 cm^{-1}.

ANSWERS TO EXAMPLES FOR FURTHER PRACTICE

Practice Example 1

The spectrum shown in Figure 7.4a is of a compound with formula C_8H_{16}. The formula tells us that there must be one C=C double bond in the molecule since the fully saturated molecule would have the formula C_8H_{18}. The band above 3000 cm^{-1} in the spectrum tells us that there are hydrogens attached to a double bond. The spectrum also shows a C=C stretching frequency at 1650 cm^{-1} and C=H deformation bands at 998 and 915 cm^{-1}, with a weak peak at 720 cm^{-1}. The C=C stretching band is intense and therefore the compound is not a *trans*-alkene with the doublc bond far from the end of the chain. The peak at 720 cm^{-1} gives similar information and tells us that we have at least four CH_2 groups. The C—H deformation bands tell us that the molecule contains the group —C=CH_2. The only possible structure is therefore oct-1-ene.

Practice Example 2

The molecule is obviously aromatic. The C—H stretching frequencies are above 3000 cm^{-1}. The peaks just below 3000 cm^{-1} are probably overtones from the very strong bands at 1530 and 1350 cm^{-1}. The bands at 850 and 700 cm^{-1} tell us that this is a mono-substituted benzene species. The two very strong bands at 1530 and 1350 cm^{-1} tell us that a NO_2 group is present. The liquid is nitrobenzene.

Practice Example 3

The strong doublet at 3380 and 3180 cm^{-1} suggests an NH_2 group. There is also a C=O stretching absorption at 1665 cm^{-1}. It looks like a primary amide and also contains a benzene ring. The molecule is actually benzamide.

Practice Example 4

Figure 7.4d shows the infrared spectrum of an unknown polymer film. The lack of CH_3 and CH_2 deformation bands in the 1500–1300 cm^{-1}

range indicates that there are no alkyl groups in the molecule. In addition, there are bands above and below 3000 cm^{-1} so the molecule must contain both aromatic and aliphatic C—H groups. Typical benzene ring absorptions are also observed: C—H stretching at 3100–3000 cm^{-1}, overtone and combination bands at 2000–1650 cm^{-1}, ring stretching at 1600–1550 cm^{-1}, ring stretching at 1500–1450 cm^{-1}, C—H in-plane bending at 1300–1000 cm^{-1}, and C—H out-of-plane bending at 900–600 cm^{-1}. The two peaks at 700 and 780 cm^{-1} indicate a mono-substituted ring. There is also a C=C stretching mode at 1639 cm^{-1} and out-of-plane bending frequencies at 998 and 915 cm^{-1}. The polymer is polystyrene. If you have had some experience with infrared spectrometers, you may recognise the spectrum, as it is commonly used as a standard for checking the stability of instruments.

Practice Example 5

Figure 7.4e shows the spectrum of a silicon-based polymer. One of the most notable features of this spectrum is a very intense broad band in the range 1100–1000 cm^{-1}, which can be readily assigned to the asymmetric Si—O—Si stretching of a siloxane. The intense band between 1300–1200 cm^{-1} indicates the presence of a Si—CH_3 group — this band is due to symmetric CH_3 deformation. Further evidence of a Si—CH_3 group is provided by bands near 1410, 850 and 760 cm^{-1}, which are due to symmetric CH_3 deformation, methyl rocking and Si—C stretching, respectively. An intense band near 1150 cm^{-1} may be assigned to Si–C_6H_5 stretching. The backbone of a silicon-based polymer must consist of a Si—O—Si backbone and we know that phenyl and methyl groups are attached to silicon. The spectrum is of α,ω-bis(trimethylsiloxy) poly(methylphenylsiloxane).

Units of Measurement

For historical reasons a number of different units of measurement have evolved to express a quantity of the same thing. In the 1960s, many international scientific bodies recommended the standardisation of names and symbols and the adoption universally of a coherent set of units — the SI units (Système Internationale d'Unités) — based on the definition of seven basic units: metre (m), kilogram (kg), second (s), ampere (A), kelvin (K), mole (mol), and candela (cd).

The earlier literature references and some of the older text books naturally use the older units. Even now many practising scientists have not adopted SI units as their working units. It is, therefore, necessary to know of the older units and to be able to interconvert these with the SI units.

In this series of texts SI units are used as standard practice. However, in areas of activity where their use has not become general practice, for example biologically based laboratories, the earlier defined units are used. This is explained in the study guide to each unit.

Table 1 shows some symbols and abbreviations commonly used in analytical chemistry, while Table 2 shows some of the alternative methods for expressing the values of physical quantities and their relationship to the values in SI units. In addition, Table 3 lists prefixes for SI units and Table 4 shows the recommended values of a selection of physical constants.

More details and definitions of other units may be found in D. H. Whiffen, *Manual of Symbols and Terminology for Physicochemical Quantities and Units*, Pergamon Press, 1979.

Table 1 *Symbols and Abbreviations Commonly Used in Analytical Chemistry*

Å	Angstrom
A_r(X)	relative atomic mass of X
A	ampere
E or U	energy
G	Gibbs free energy (function)
H	enthalpy
J	joule
K	kelvin (273.15 + t°C)
K	equilibrium constant (with subscripts p, c, therm, etc.)
K_a,K_b	acid and base ionisation constants
M_r(X)	relative molecular mass of X
N	newton (SI unit of force)
P	total pressure
s	standard deviation
T	temperature/K
V	volume
V	volt (J A^{-1} s^{-1})
a, a(A)	activity, activity of A
c	concentration/mol dm^{-3}
e	electron
g	gram
i	current
s	second
t	temperature/°C
bp	boiling point
fp	freezing point
mp	melting point
$\approx$	approximately equal to
$<$	less than
$>$	greater than
e, exp(x)	exponential of x
ln x	natural logarithm of x; $\ln x = 2.303 \log x$
log x	common logarithm of x to base 10

Table 2 *Summary of Alternative Methods of Expressing Physical Quantities*

(1) **Mass (SI unit: kg)**

$g = 10^{-3}\,kg$
$mg = 10^{-3}\,g = 10^{-6}\,kg$
$\mu g = 10^{-6}\,g = 10^{-9}\,kg$

(2) **Length (SI unit: m)**

$cm = 10^{-2}\,m$
$Å = 10^{-10}\,m$
$nm = 10^{-9}\,m = 10\,Å$
$pm = 10^{-12}\,m = 10^{-2}\,Å$

(3) **Volume (SI unit: m^3)**

$l = dm^3 = 10^{-3}\,m^3$
$ml = cm^3 = 10^{-6}\,m^3$
$\mu l = 10^{-3}\,cm^3$

(4) **Concentration (SI unit: mol m^{-3})**

$M = mol\,l^{-1} = mol\,dm^{-3} = 10^3\,mol\,m^{-3}$
$mg\,l^{-1} = \mu g\,cm^{-3} = ppm = 10^{-3}\,g\,dm^{-3}$
$\mu g\,g^{-1} = ppm = 10^{-6}\,g\,g^{-1}$
$ng\,cm^{-3} = ppb = 10^{-6}\,g\,dm^{-3}$
$pg\,g^{-1} = ppt = 10^{-12}\,g\,g^{-1}$
$mg\% = 10^{-2}\,g\,dm^{-3}$
$\mu g\% = 10^{-5}\,g\,dm^{-3}$

(5) **Pressure (SI unit: $N\,m^{-2} = kg\,m^{-1}\,s^{-2}$)**

$Pa = N\,m^{-2}$
$atm = 101\,325\,N\,m^{-2}$
$bar = 10^5\,N\,m^{-2}$
$torr = mmHg = 133.322\,N\,m^{-2}$

(6) **Energy (SI unit: $J = kg\,m^2\,s^{-2}$)**

$cal = 4.184\,J$
$erg = 10^{-7}\,J$
$eV = 1.602 \times 10^{-19}\,J$

Table 3 *Prefixes for SI Units*

Fraction	**Prefix**	**Symbol**
10^{-1}	deci	d
10^{-2}	centi	c
10^{-3}	milli	m
10^{-6}	micro	μ
10^{-9}	nano	n
10^{-12}	pico	p
10^{-15}	femto	f
10^{-18}	atto	a
Multiple	**Prefix**	**Symbol**
10	deka	da
10^{2}	hecto	h
10^{3}	kilo	k
10^{6}	mega	M
10^{9}	giga	G
10^{12}	tera	T
10^{15}	peta	P
10^{18}	exa	E

Table 4 *Recommended Values of Physical Constants*

Physical constant	Symbol	Value
acceleration due to gravity	g	$9.81\ \mathrm{m\,s^{-2}}$
Avogadro constant	N_A	$6.022\,14 \times 10^{23}\ \mathrm{mol^{-1}}$
Boltzmann constant	k	$1.380\,66 \times 10^{-23}\ \mathrm{J\,K^{-1}}$
charge-to-mass ratio	e/m	$1.758\,796 \times 10^{11}\ \mathrm{C\,kg^{-1}}$
electronic charge	e	$1.602\,18 \times 10^{-19}\ \mathrm{C}$
Faraday constant	F	$9.648\,46 \times 10^{4}\ \mathrm{C\,mol^{-1}}$
gas constant	R	$8.314\ \mathrm{J\,K^{-1}\,mol^{-1}}$
`ice-point' temperature	T_{ice}	273.150 K, exactly
molar volume of ideal gas (stp)	V_m	$2.241\,38 \times 10^{-2}\ \mathrm{m^3\,mol^{-1}}$
permittivity of a vacuum	ϵ_0	$8.854\,188 \times 10^{-12}$ $\mathrm{kg^{-1}\,m^{-3}\,s^4\,A^2}$ $(\mathrm{F\,m^{-1}})$
Planck constant	h	$6.626\,08 \times 10^{-34}\ \mathrm{J\,s}$
standard atmosphere pressure	p	$101\,325\ \mathrm{N\,m^{-2}}$, exactly
atomic mass unit (amu)	m_u	$1.660\,54 \times 10^{-27}\ \mathrm{kg}$
speed of light in a vacuum	c	$2.997\,925 \times 10^{8}\ \mathrm{m\,s^{-1}}$

Index